PHILOSOPHY OF MATHEMATICS

In his long-awaited new edition of *Philosophy of Mathematics*, James Robert Brown tackles important new as well as enduring questions in the mathematical sciences. Can pictures go beyond being merely suggestive and actually prove anything? Are mathematical results certain? Are experiments of any real value?

This clear and engaging book takes a unique approach, encompassing non-standard topics such as the role of visual reasoning, the importance of notation, and the place of computers in mathematics, as well as traditional topics such as formalism, Platonism, and constructivism. The combination of topics and clarity of presentation make it suitable for beginners and experts alike. The revised and updated second edition of *Philosophy of Mathematics* contains more examples, suggestions for further reading, and expanded material on several topics including a novel approach to the continuum hypothesis.

Praise for the first edition:

'This book is a breath of fresh air for undergraduate philosophy of mathematics. Very accessible and even entertaining, Brown explains most of the issues without technicalities.'

Janet Folina, *Macalester College*

'a wonderful introduction to the philosophy of mathematics. It's lively, accessible, and, above all, a terrific read. It would make an ideal text for an undergraduate course on the philosophy of mathematics; indeed, I recommend it to anyone interested in the philosophy of mathematics – even specialists in the area can learn from this book.'

Mark Colyvan, *University of Sydney*

James Robert Brown is Professor of Philosophy at the University of Toronto, Canada.

Routledge Contemporary Introductions to Philosophy
Series editor: Paul K Moser
Loyola University of Chicago

This innovative, well-structured series is for students who have already done an introductory course in philosophy. Each book introduces a core general subject in contemporary philosophy and offers students an accessible but substantial transition from introductory to higher-level college work in that subject. The series is accessible to non-specialists and each book clearly motivates and expounds the problems and positions introduced. An orientating chapter briefly introduces its topic and reminds readers of any crucial material they need to have retained from a typical introductory course. Considerable attention is given to explaining the central philosophical problems of a subject and the main competing solutions and arguments for those solutions. The primary aim is to educate students in the main problems, positions and arguments of contemporary philosophy rather than to convince students of a single position.

Classical Philosophy
Christopher Shields

Epistemology
Second Edition
Robert Audi

Ethics
Harry Gensler

Metaphysics
Second Edition
Michael J. Loux

Philosophy of Art
Noël Carroll

Philosophy of Language
Willam G. Lycan

Philosophy of Mathematics:
A Contemporary Introduction to
the World of Proofs and Pictures
Second Edition
James R. Brown

Philosophy of Mind
Second Edition
John Heil

Philosophy of Religion
Keith E. Yandell

Philosophy of Science
Alex Rosenberg

Social and Political Philosophy
John Christman

Philosophy of Psychology
José Bermudez

Classical Modern Philosophy
Jeffrey Tlumak

Philosophy of Biology
Alex Rosenberg and Daniel
W. McShea

PHILOSOPHY OF MATHEMATICS

A Contemporary Introduction to the World of Proofs and Pictures

Second Edition

James Robert Brown

Routledge
Taylor & Francis Group

NEW YORK AND LONDON

First edition published 1999
by Routledge

This edition first published 2008
by Routledge
270 Madison Ave, New York, NY 10016

Simultaneously published in the UK
by Routledge
2 Park Square, Milton Park, Abingdon, Oxon OX14 4RN

*Routledge is an imprint of the Taylor & Francis Group,
an informa business*

First edition © 1999 James Robert Brown

Second edition © 2008 Taylor & Francis

Typeset in Times by
Integra Software Services Pvt. Ltd, Pondicherry, India
Printed and bound in the United States of America
on acid-free paper by Walsworth Publishing Company,
Marceline, MO

Library of Congress Cataloging in Publication Data
Brown, James Robert.
Philosophy of mathematics: a contemporary introduction to the world
of proofs and pictures / James Robert Brown. -- 2nd ed.
p. cm.
Includes bibliographical references and index.
ISBN 978-0-415-96048-9 (hardback : alk. paper) --
ISBN 978-0-415-96047-2 (pbk. : alk. paper) --
ISBN 978-0-203-93296-4 (ebook)
1. Mathematics--Philosophy.   I. Title.
QA8.4.B76 2008
510.1--dc22
2007042483

ISBN10: 0–415–96048–7 (hbk)
ISBN10: 0–415–96047–9 (pbk)
ISBN10: 0–203–93296–X (ebk)

ISBN13: 978–0–415–96048–9 (hbk)
ISBN13: 978–0–415–96047–2 (pbk)
ISBN13: 978–0–203–93296–4 (ebk)

For Elizabeth and Stephen

Contents

Preface and Acknowledgements

A philosopher who has nothing to do with geometry is only half a philosopher, and a mathematician with no element of philosophy in him is only half a mathematician. These disciplines have estranged themselves from one another to the detriment of both.

Frege

A heavy warning used to be given that pictures are not rigorous; this has never had its bluff called and has permanently frightened its victims into playing for safety. Some pictures, of course, are not rigorous, but I should say most are (and I use them whenever possible myself).

Littlewood

There are a number of ways in which this book could fail. It has several goals, some of them pedagogical. One of these goals is to introduce readers to the philosophy of mathematics. In my attempt to avoid failure here I've included chapters on traditional points of view, such as formalism and constructivism, as well as Platonism. And since I'm aiming at a broad audience, I've taken pains to explain philosophical notions that many readers may encounter for the first time. I've also given lots of detailed mathematical examples for the sake of those who lack a technical background. It's been my experience that there is a huge number of students who come to philosophy from a humanities background wanting to know a bit about the sciences, and when they are properly introduced they find that their appetites for mathematics become insatiable. I'd be delighted to stimulate a few readers in this way.

If we taught philosophy today in a way that reflected its history, the current curriculum would be overwhelmed with the philosophy of mathematics. Think of these great philosophers and how important mathematics is to their thought: Plato, Descartes, Leibniz, Kant, Frege, Russell, Wittgenstein, Quine, Putnam, and so many others. And interest in the nature of mathematics is not confined to the so-called analytic stream of philosophy; it also looms large in the work of Husserl and Lonergin, central figures in, respectively, the continental and Thomistic philosophical traditions. Anyone sincerely interested in philosophy must be interested in the nature of mathematics, and I hope to show why. As for those who persist in thinking otherwise – let them burn in hell.

This book could also fail in a second, more important aim, which is to introduce some of the newer issues in the philosophy of mathematics, namely those associated with computers, 'experimentation', and especially with visualization. Traditional issues remain fascinating and unresolved; philosophers and mathematicians alike continue to work on them. (Even logicism, the view that mathematics is really just logic, is making a partial comeback.) But if there are living philosophical issues for working mathematicians, they have to do with the role of computers and computer graphics and the role of physics within mathematics. Some consider the use of computers a glorious revolution – others think it a fraud. Some are thrilled at the new relations with physics – others fear the fate of rigour. Current battles are just as lively as those between Russell and Poincaré early in the twentieth century or between Hilbert and Brouwer in the 1920s and 1930s. And philosophers should know about them. This book would be a failure if something of the content of the issues and the spirit of current debates is not conveyed.

Finally, I could fail in my attempt to argue for Platonism, in general, and, for a Platonistic account of how (some) pictures work, in particular. Mathematicians are instinctively realists; but when forced to think about the details of this realism, they often become uneasy. Philosophers, aware of the bizarreness of abstract objects, are already wary of mathematical realism. But still, most people are somewhat sympathetic to Platonism in mathematics, tolerant to an extent that they wouldn't tolerate, say, Platonism in physics or in ethics. My case for Platonism will meet with at least mild resistance, but this is nothing compared with the hostility that will greet my account of how picture-proofs work. On this last point I expect to fail completely in winning over readers. But I will be somewhat mollified if it is generally admitted that the problems this work raises and addresses are truly wonderful, worthy of wide attention.

I've had a great deal of help from a great many people in a great many ways. Some are long-time colleagues with whom I've been arguing these issues longer than we care to remember. Some are students subjected to earlier drafts. Some listened to an argument. Some read a chapter. Some worked carefully through the whole of an earlier draft. For their help in whatever form, enormous thanks go to: Peter Apostoli, Michael Ashooh, John Bell, Gordon Belot, Alexander Bird, Elizabeth East, Danny Goldstick, Ian Hacking, Michael Hallett, Sarah Hoffman, Andrew Irvine, Loki Jorgenson, Bernard Katz, Margery Konan, Hugh Lehman, Mary Leng, Dennis Lomas, Ken Manders, James McAllister, Patrick

Moran, Margaret Morrison, Joshua Mozersky, Bill Newton-Smith, Calvin Normore, John Norton, Kathleen Okruhlik, David Papineau, Fred Portoraro, Bill Seager, Zvonimir Šikić, Spas Spassov, Mary Tiles, Jacek Urbaniec, Alasdair Urquhart, and Katherine Van Uum. I've read chapters at various conferences and to various philosophy departments, and in every case I've much benefited from audience comments though, more often than not, I don't know whom to thank. Some of the material was presented as the Matchette Lectures at Purdue University; I'm especially grateful to Martin and Pat Curd and their Philosophy Department for arranging what was for me a great week.

Chapters 3, 4 and 7 are revised from earlier articles. I thank Oxford University Press and Kluwer Academic Publishers for their kind permission to use this material.

Finally, I'm very grateful to SSHRC for its support.

Preface to the Second Edition

This edition differs from the first in several respects. I have made numerous minor modifications and corrections throughout. I'm very grateful to all those who pointed out mistakes or unclear passages. In a few places I have over-hauled whole paragraphs. And I have added a brief 'Further Reading' section to the end of each chapter.

The biggest change comes in the form a new chapter on the continuum hypothesis (Chapter 11). This has been one of the great problems of mathematics for more than a century. It was shown by Gödel and Cohen to be independent, hence neither provable nor refutable, given the other axioms of set theory. Realists claim it has a truth-value, nevertheless. Could we ever come to know what that truth value is? Christopher Freiling may have refuted it by means of a thought experiment. This is certainly not the usual way of doing mathematics, but if it works – and I'm inclined to think it does – then it wonderfully illustrates the power of new techniques, such as visual reasoning.

I expect this new chapter to be controversial – indeed, I hope it is. Even if it fails as a refutation of the continuum hypothesis, I'll be gratified if it provokes deeper reflection on the cluster of issues associated with the continuum hypothesis and with visual thinking in mathematics, in general. That remains a main aim throughout the whole book.

I'm very pleased with the reception of the first edition. The reviewers were overly generous (a fault I'm happy to pardon). Students seemed to get something out of it and even enjoyed doing so. Experts found things to contest. I couldn't reasonably ask for more – but I will. If I had any disappointment, it would concern the more esoteric topics. For instance, reviewers often remarked enthusiastically on the potential interest and importance of topics such as notation. But as far as I know, the theme has not been further explored. I suppose all I can do is once again urge others to take up the matter. The old topics such as constructivism and formalism remain interesting, and so are the newer ones such as indispensability and structuralism. But there is a goldmine waiting for us all in the issues of visualization, notation, computer simulation, and mathematical thought experimentation. That's where the future lies.

A number of people need to be thanked. I reiterate my thanks to those who helped with the first edition. Some of these and some others were very helpful this time, as well. In particular, thanks go to Ken Manders and Louis Levin for finding mistakes and typos, to Zvonimir Šikić for critical comments on graphs, and Chris Freiling for comments on my account of his work on the continuum. I also learned a great deal from reviewers of the first edition. A list of these can be found at http://www.chass.utoronto.ca/~jrbrown/NOTES.Philosophy_of_Mathematics. Any additional comments, corrections, and reviews of this second edition will also be listed there.

CHAPTER 1
Introduction: The Mathematical Image

Let's begin with a nice example, the proof that there are infinitely many prime numbers. If asked for a typical bit of real mathematics, your friendly neighbourhood mathematician is as likely to give this example as any. First, we need to know that some numbers, called 'composite', can be divided without remainder or broken into factors (e.g. $6 = 2 \times 3$, $561 = 3 \times 11 \times 17$), while other numbers, called 'prime', cannot (e.g. 2, 3, 5, 7, 11, 13, 17, . . .). Now we can ask: How many primes are there? The answer is at least as old as Euclid and is contained in the following.

Theorem: There are infinitely many prime numbers.

Proof: Suppose, contrary to the theorem, that there is only a finite number of primes. Thus, there will be a largest which we can call p. Now define a number n as 1 plus the product of all the primes:

$$n = (2 \times 3 \times 5 \times 7 \times 11 \times \ldots \times p) + 1$$

Is n itself prime or composite? If it is prime then our original supposition is false, since n is larger than the supposed largest prime p. So now let's consider it composite. This means that it must be divisible (without remainder) by prime numbers. However, none of the primes up to p will divide n (since we would always have remainder 1), so any number which does divide n must be greater than p. This means that there is a prime number greater than p after all. Thus, whether n is prime or composite, our supposition that there is a largest prime number is false. Therefore, the set of prime numbers is infinite.

The proof is elegant and the result profound. Still, it is typical mathematics; so, it's a good example to reflect upon. In doing so, we will begin to see the elements of *the mathematical image*, the standard conception of what mathematics is. Let's begin a list of some commonly accepted aspects. By 'commonly accepted' I mean that they would be accepted by most working mathematicians, by most educated people, and probably by most philosophers of mathematics, as well. In listing them as part of the common mathematical image we need not endorse them. Later we may even come to reject some of them – I certainly will. With this caution in mind, let's begin to outline the standard conception of mathematics.

Certainty   The theorem proving the infinitude of primes seems established beyond a doubt. The natural sciences can't give us anything like this. In spite of its wonderful accomplishments, Newtonian physics has been overturned in favour of quantum mechanics and relativity. And no one today would bet too heavily on the longevity of current theories. Mathematics, by contrast, seems the one and only place where we humans can be absolutely sure we got it right.

Objectivity   Whoever first thought of this theorem and its proof made a great discovery. There are other things we might be certain of, but they aren't discoveries: 'Bishops move diagonally.' This is a chess rule; it wasn't discovered; it was invented. It is certain, but its certainty stems from our resolution to play the game of chess that way. Another way of describing the situation is by saying that our theorem is an objective truth, not a convention. Yet a third way of making the same point is by saying that Martian mathematics is like ours, while their games might be quite different.

Proof is essential   With a proof, the result is certain; without it, belief should be suspended. That might be putting it a bit too strongly. Sometimes mathematicians believe mathematical propositions even though they lack a proof. Perhaps we should say that without a proof a mathematical proposition is not justified and should not be used to derive other mathematical propositions. Goldbach's conjecture is an example. It says that every even number is the sum of two primes. And there is lots of evidence for it, e.g. $4 = 2 + 2, 6 = 3 + 3, 8 = 3 + 5, 10 = 5 + 5, 12 = 7 + 5$, and so on. It's been checked into the billions without a counter-example. Biologists don't hesitate to conclude that all ravens are black based on this sort of evidence; but mathematicians (while they might believe that Goldbach's conjecture is true) won't call it a theorem and won't use it to establish other theorems – not without a proof.

   Let's look at a second example, another classic, the Pythagorean theorem. The proof below is modern, not Euclid's.

   Theorem: In any right-angled triangle, the square of the hypotenuse is equal to the sum of the squares on the other two sides.

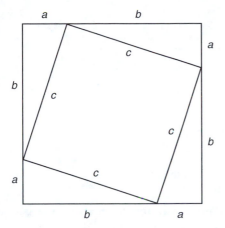

Figure 1.1

Proof: Consider two square figures, the smaller placed in the larger, making four copies of a right-angled triangle $\triangle abc$ (Figure 1.1). We want to prove that $c^2 = a^2 + b^2$.

The area of the outer square $= (a + b)^2 = c^2 + 4 \times$ (area of $\triangle abc$) $= c^2 + 2ab$, since the area of each copy of $\triangle abc$ is $\frac{1}{2} ab$. From algebra we have $(a + b)^2 = a^2 + 2ab + b^2$. Subtracting $2ab$ from each, we conclude $c^2 = a^2 + b^2$.

This brings out another feature of the received view of mathematics.

Diagrams   There are no illustrations or pictures in the proofs of most theorems. In some there are, but these are merely a psychological aide. The diagram helps us to understand the theorem and to follow the proof – nothing more. The proof of the Pythagorean theorem or any other is the verbal/symbolic argument. Pictures can never play the role of a real proof.

Remember, in saying this I'm not endorsing these elements of the mathematical image, but merely exhibiting them. Some of these I think right, others, including this one about pictures, quite wrong. Readers might like to form their own tentative opinions as we look at these examples.

Misleading diagrams   Pictures, at best, are mere psychological aids; at worst they mislead us – often badly. Consider the infinite series

$$\sum_{n=1}^{\infty} \frac{1}{n^2} = 1 + \frac{1}{4} + \frac{1}{9} + \frac{1}{16} + \ldots$$

which we can illustrate with a picture (Figure 1.2):

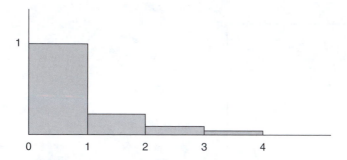

Figure 1.2 Shaded blocks correspond to terms in the series

The sum of this series is $\pi^2/6 = 1.6449\ldots$ In the picture, the sum is equal to the shaded area. Let's suppose we paint the area and that this takes *one* can of paint.

Next consider the so-called harmonic series

$$\sum_{n=1}^{\infty} \frac{1}{n} = 1 + \frac{1}{2} + \frac{1}{3} + \frac{1}{4} + \ldots$$

Here's the corresponding picture (Figure 1.3):

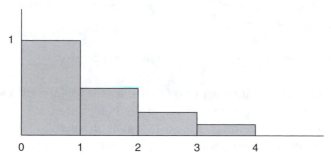

Figure 1.3

The steps keep getting smaller and smaller, just as in the earlier case, though not quite so fast. How big is the shaded area? Or rather, how much paint will be required to cover the shaded area? Comparing the two pictures, one would be tempted to say that it should require only slightly more – perhaps two or three cans of paint at most. Alas, such a guess couldn't be further off the mark. In fact, there isn't enough paint in the entire universe to cover the shaded area – it's infinite. The proof goes as follows. As we write out the series, we can group the terms:

$$\underbrace{\frac{1}{1}} + \underbrace{\frac{1}{2} + \frac{1}{3}} + \underbrace{\frac{1}{4} + \frac{1}{5} + \frac{1}{6} + \frac{1}{7}} + \underbrace{\frac{1}{8} + \frac{1}{9} + \ldots}$$

The size of the first group is obviously 1. In the second group the terms are between $\frac{1}{2}$ and $\frac{1}{4}$, so the size is between $2 \times \frac{1}{2}$ and $2 \times \frac{1}{4}$, that is, between $\frac{1}{2}$ and 1. In the next grouping of four, all terms are bigger than 1/8, so the sum is again between $\frac{1}{2}$ and 1. The same holds for the next group of 8 terms; it, too, has a sum between $\frac{1}{2}$ and 1. Clearly, there are infinitely many such groupings, each with a sum greater than $\frac{1}{2}$. When we add them all together, the total size is infinite. It would take more paint than the universe contains to cover it all. Yet, the picture doesn't give us an inkling of this startling result.

One of the most famous results of antiquity still amazes; it is the proof of the irrationality of the square root of 2. A rational number is a ratio, a fraction, such as 3/4 or 6937/528, which is composed of whole numbers. $\sqrt{9} = 3$ is rational and so is $\sqrt{(9/16)} = 3/4$; but $\sqrt{2}$ is not rational as the following theorem shows.

Theorem: The square root of 2 is not a rational number.

Proof: Suppose, contrary to the theorem, that $\sqrt{2}$ is rational, i.e. suppose that there are integers p and q such that $\sqrt{2} = p/q$. Or equivalently, $2 = (p/q)^2 = p^2/q^2$. Let us further assume that p/q is in lowest terms. (Note that $3/4 = 9/12 = 21/28$, but only the first expression is in lowest terms.)

Rearranging the above expression, we have $p^2 = 2q^2$. Thus, p^2 is even (because 2 is a factor of the right side). Hence, p is even (since the square of an odd number is odd). So it follows that $p = 2r$, for some number r. From this we get $2q^2 = p^2 = (2r)^2 = 4r^2$. Thus, $q^2 = 2r^2$, which implies that q^2 is even, and hence that q is even.

Now we have the result that both p and q are even, hence both divisible by 2, and so, not in lowest terms as was earlier supposed. Thus, we have arrived at the absurdity that p/q both *is* and *is not* in lowest terms. Therefore, our initial assumption that $\sqrt{2}$ is a rational number is false.

Classical logic   Notice the structure of the proof of the irrationality of $\sqrt{2}$. We made a supposition. We derived a contradiction from this, showing the supposition is false. Then we concluded that the negation of the supposition is true. The logical principles behind this are: first, no proposition is both true and false (non-contradiction) and second, if a proposition is false, then its negation is true (excluded middle). Classical logic is a working tool of mathematics. Without this tool, much of traditional mathematics would crumble.

Strictly speaking, the proof of the irrationality of $\sqrt{2}$ is acceptable to constructive mathematicians, even though they deny the general legitimacy of classical logic. The issue will come up in more detail in a later chapter. The proof just given nicely illustrates *reduction ad absurdum* reasoning. It is also one of the all time great results, which everyone should know as a matter of general culture, just as everyone should know *Hamlet*. This is my excuse for using an imperfect example to make the point about classical logic.

Sense experience   All measurement in the physical world works perfectly well with rational numbers. Letting the standard metre stick be our unit, we can measure any length with whatever desired accuracy our technical abilities will allow; but the accuracy will always be to some rational number (some fraction of a metre). In other words, we could not discover irrational numbers or incommensurable segments (i.e. lengths which are not ratios of integers) by physical measurement. It is sometimes said that we learn $2 + 2 = 4$ by counting apples and the like. Perhaps experience plays a role in grasping the elements of the natural numbers. But the discovery of the irrationality of $\sqrt{2}$ was an intellectual achievement, not at all connected to sense experience.

Cumulative history   The natural sciences have revolutions. Cherished beliefs get tossed out. But a mathematical result, once proven, lasts forever. There are mathematical revolutions in the sense of spectacular results which yield new methods to work with and which focus attention in a new field – but no theorem is ever overturned. The mathematical examples I have so far discussed all pre-date Ptolemaic astronomy, Newtonian mechanics, Christianity and capitalism; and no doubt they will outlive them all. They are permanent additions to humanity's collection of glorious accomplishments.

Computer proofs   Computers have recently played a dramatic role in mathematics. One of the most celebrated results has to do with map colouring. How many colours are needed to insure that no adjacent countries are the same colour?

   Theorem: Every map is four-colourable.

I won't even try to sketch the proof of this theorem. Suffice it to say that a computer was set the task of checking a very large number of cases. After a great many hours of work, it concluded that there are no counter-examples to the theorem: every map can be coloured with four colours. Thus, the theorem was established.

   It's commonplace to use a hand calculator to do grades or determine our finances. We could do any of these by hand. The little gadget is a big time saver and often vastly more accurate than our efforts. Otherwise, there's really nothing new going on. Similarly, when a supercomputer tackles a big problem and spends hours on its solution, there is nothing new going on there either. Computers do what we do, only faster and perhaps more accurately. Mathematics hasn't changed because of the introduction of computers. A proof is still a proof, and that's the one and only thing that matters.

Solving problems   There are lots of things we might ask, but have little chance of answering: 'Does God exist?' 'Who makes the best pizza?' These seem perfectly meaningful questions, but the chances of finding answers seems hopeless.

By contrast, it seems that every mathematical question can be answered and every problem solved. Is every even number (greater than 2) equal to the sum of two primes? We don't know now, but that's because we've been too stupid so far. Yet we are not condemned to ignorance about Goldbach's conjecture the way we are about the home of the best pizza. It's the sort of question that we should be able to answer, and in the long run we will.

Having said this, a major qualification is in order. In fact, we may have to withdraw the claim. So far, in listing the elements of the mathematical image we've made no distinction among mathematicians, philosophers and the general public. But at this point we may need to distinguish. Recent results such as Gödel's incompleteness theorem, the independence of the continuum hypothesis and others have led many mathematicians and philosophers of mathematics to believe that there are problems which are unsolvable in principle. The pessimistic principle would seem to be part of the mathematical image.

Well, enough of this. We've looked at several notions that are very widely shared and, whether we endorse them or not, they seem part of the common conception of mathematics. In sum, these are a few of the ingredients in the mathematical image:

(1) *Mathematical results are certain*
(2) *Mathematics is objective*
(3) *Proofs are essential*
(4) *Diagrams are psychologically useful, but prove nothing*
(5) *Diagrams can even be misleading*
(6) *Mathematics is wedded to classical logic*
(7) *Mathematics is independent of sense experience*
(8) *The history of mathematics is cumulative*
(9) *Computer proofs are merely long and complicated regular proofs*
(10) *Some mathematical problems are unsolvable in principle*

More could be added, but this is grist enough for our mill. Here we have the standard conception of mathematics shared by most mathematicians and non-mathematicians, including most philosophers. Yet not everyone accepts this picture. Each of these points has its several critics. Some deny that mathematics was ever certain and others say that, given the modern computer, we ought to abandon the ideal of certainty in favour of much more experimental mathematics. Some deny the objectivity of mathematics, claiming that it is a human invention after all, adding that though it's a game like chess, it is the greatest game ever played. Some deny that classical logic is indeed the right tool for mathematical inference, claiming that there are indeterminate (neither true nor false) mathematical propositions. And, finally, some would claim great virtues for pictures as proofs, far beyond their present lowly status.

We'll look at a number of issues in the philosophy of mathematics, some traditional, some current, and we'll see how much of the mathematical image endures this scrutiny. Don't be surprised should you come to abandon at least some of it. I will.

Further Reading

Many come to the philosophy of mathematics before a serious encounter with mathematics itself. If you're looking for a good place to get your feet wet, try an old classic, by Courant, Robins, and Stewart, *What is Mathematics*? If you're trying to teach yourself mathematics using standard textbooks, then I strongly urge reading popular books, as well. Rough analogies, anecdotes, and even gossip are an important part of any mathematical education. Biographies are important, too. For a collection of brief biographies of several contemporaries, try Albers and Alexanderson (eds) *Mathematical People*. There are several introductory books in the philosophy of mathematics. Shapiro, *Talking About Mathematics* is particularly nice; it covers traditional topics and Shapiro's own 'structuralism'.

CHAPTER 2
Platonism

What's the greatest discovery in the history of thought? Of course, it's a silly question – but it won't stop me from suggesting an answer. It's Plato's discovery of abstract objects. Most scientists, and indeed most philosophers, would scoff at this. Philosophers admire Plato as one of the greats, but think of his doctrine of the heavenly forms as belonging in a museum. Mathematicians, on the other hand, are at least slightly sympathetic. Working day-in and day-out with primes, polynomials and principal fibre bundles, they have come to think of these entities as having a life of their own. Could this be only a visceral reaction to an illusion? Perhaps, but I doubt it. The case for Platonism, however, needs to made carefully. Let's begin with a glance at the past.

The Original Platonist

We notice a similarity among various apples and casually say, 'There is something they have in common.' But what could this *something* they have in common be? Should we even take such a question literally? Plato did and said the common thing is *the form of an apple*. The form is a perfect apple, or perhaps a kind of blueprint. The actual apples we encounter are copies of the form; some are better copies than others. A dog is a dog in so far as it 'participates' in *the form of a dog*, and an action is morally just in so far as it participates in *the form of justice*.

   How do we know about the forms? Our immortal souls once resided in heaven and in this earlier life gazed directly upon the forms. But being born into this world was hard on our memories; we forgot everything. Thus, according to Plato, what we call learning is actually recollection. And so, the proper way to teach is the so-called Socratic method of questioning, which does not simply state the facts to us, but instead helps us to remember what we already know.

The example of the slave-boy in the *Meno* dramatically illustrates Plato's point. After being assured that the slave-boy has had no mathematical training, Socrates, through a clever sequence of questions, gets him to double the square, which, for a novice, is a rather challenging geometric construction. Not only does the slave-boy do it but, after a false start, he recognizes that he has finally done it correctly. Plato's moral is that the slave-boy *already* knew how to double the square, and Socrates, the self-described 'mid-wife', simply helped him in bringing out what was already there.

Plato's theory is both wonderful and preposterous. It's wonderful because of its tremendous scope. It explains what all apples have in common, what makes a moral act moral, how we acquire knowledge and, above all, it tells us what mathematics is. This last feature especially rings true – even if nothing else about Platonism does. When we talk about circles, for example, we don't seem to be talking about any particular figure on the blackboard. Those are only approximations. We're talking about a perfect circle, something which exists nowhere in the physical world. At this point it's completely natural to feel drawn towards Plato's realm of eternal forms. And many find the tug irresistible.[1]

But the theory is also preposterous. What possible sense can we make of abstract entities? Immortal souls? Recollection? Philosophers who think Plato's forms belong in a museum are, for the most part, right. However, contemporary Platonism needn't embrace all of this. The essential ingredient is *the existence and accessibility of the forms themselves, in particular the mathematical forms*, if not the others such as tallness and justice. That's all that current mathematical Platonism is concerned with. But even this is a huge assumption. Spelling it out and making it plausible is no easy task.

Some Recent Views

Many of the greatest mathematicians and logicians have been gung-ho Platonists. Let's sample some views, starting with Gottlob Frege, arguably the greatest logician and (after Plato) the greatest philosopher of mathematics of all time. My aim in citing Frege and other luminaries is twofold: in part to get a feel for Platonism and in part to appeal shamelessly to authority – if these smart guys believe it, we should at least take it seriously.

Frege distinguishes among our *ideas* (which are psychological entities), *thoughts*, as he calls them (which are the content of our ideas), and the sentences we use to express them (which, as he says, are *things of the outer world*, physical entities such as trees, electrons, ink marks or sound waves). Frege's *thoughts* are Platonic entities.

So the result seems to be: thoughts are neither things of the outer world nor ideas.

A third realm must be recognized. What belongs to this corresponds with ideas, in that it cannot be perceived by the senses, but with things, in that it needs no bearer to the contents of whose consciousness to belong. Thus the thought, for example, which we express in the Pythagorean theorem is timelessly true, true independently of whether anyone takes it to be true. It needs no bearer. It is not true for the first time when it is discovered, but is like a planet which, already before anyone has seen it, has been in interaction with other planets.

(Frege 1918: 523)

G.H. Hardy was one of the century's great mathematicians, famous, among other things, for his collaborations with Littlewood and with Ramanujan. His classic essay, 'Mathematical Proof', contains such Platonistic pronouncements as these:

It seems to me that no philosophy can possibly be sympathetic to the mathematician which does not admit, in one manner or another, the immutable and unconditional validity of mathematical truth. Mathematical theorems are true or false; their truth or falsity is absolute and independent of our knowledge of them. In *some* sense, mathematical truth is part of objective reality.

(Hardy 1929: 4)

I have myself always thought of a mathematician as in the first instance an *observer*, a man who gazes at a distant range of mountains and notes down his observations. His object is simply to distinguish clearly and notify to others as many different peaks as he can. There are some peaks which he can distinguish easily, while others are less clear. He sees A sharply, while of B he can obtain only transitory glimpses. At last he makes out a ridge which leads from A, and following it to its end he discovers that it culminates in B. B is now fixed in his vision, and from this point he can proceed to further discoveries. In other cases perhaps he can distinguish a ridge which vanishes in the distance, and conjectures that it leads to a peak in the clouds or below the horizon. But when he sees a peak he believes that it is there simply because he sees it. If he wishes someone else to see it, he *points to it*, either directly or through the chain of summits which led him to recognize it himself. When his pupil also sees it, the research, the argument, the *proof* is finished. The analogy is a rough one, but I am sure that it is not altogether misleading. If we were to push it to its extreme we should be led to a rather paradoxical conclusion; that there is, strictly, no such thing as mathematical proof; that we can, in the last analysis, do nothing but *point*; that proofs are what Littlewood and I call *gas*, rhetorical flourishes designed to affect psychology, pictures on the board in the lecture, devices to stimulate the imagination of pupils.

(Hardy 1929: 18)

Kurt Gödel, who achieved some of the most spectacular results in the foundations of mathematics, was also the most famous and influential Platonist of recent times. He declares that 'classes and concepts may . . . be conceived as real objects . . . existing independently of our definitions and constructions'. He draws an analogy between mathematics and physics:

> the assumption of such objects is quite as legitimate as the assumption of physical bodies and there is quite as much reason to believe in their existence. They are in the same sense necessary to obtain a satisfactory system of mathematics as physical bodies are necessary for a satisfactory theory of our sense perceptions.
>
> (Gödel 1944: 456f.)

> despite their remoteness from sense experience, we do have something like a perception also of the objects of set theory, as is seen from the fact that the axioms force themselves upon us as being true. I don't see any reason why we should have any less confidence in this kind of perception, i.e. in mathematical intuition, than in sense perception.
>
> (Gödel 1947: 484)

What is Platonism?

There are several points we can glean from the above remarks and from other writings of various Platonists. They form the core of Platonism.

(1) *Mathematical objects are perfectly real and exist independently of us*
Mathematical objects are no different than everyday objects (pine trees) or the exotic entities of science (positrons). We don't in any way create them; we discover them. And our theorems try to correctly describe them. Any well-formed sentence of mathematics is true or is false, and what makes it so are the objects to which the sentence refers. The truth of these propositions has nothing to do with us; it does not rest on the structure of our minds, nor on the way we use language, nor on the way we verify our conjectures.

This outlook gives Platonism a great advantage over its rivals who must do lots of fancy footwork to account for mathematical truth. Formalism (as we'll see in a later chapter) identifies truth with proof, but, as a result of Gödel's theorem, runs into apparent truths that cannot be formally proven. Thus the identification of truth with proof is broken. And constructivists (as we'll also see below) link truth with constructive proof, but necessarily lack constructions for many highly desirable results of classical mathematics, making their account of mathematical truth rather implausible.

Platonism and standard semantics (as it is often called), go hand-in-hand. Standard semantics is just what you think it is. Let us suppose the sentence 'Mary loves ice cream' is true. What makes it so? In answering such a question, we'd say 'Mary' refers to the person Mary, 'ice cream' to the substance, and 'loves' refers to a particular relation which holds between Mary and ice cream. It follows rather trivially from this that Mary exists. If she didn't, then 'Mary loves ice cream' couldn't be true, any more than 'Phlogiston is released on burning' could be true when phlogiston does not exist.

The same semantical considerations imply Platonism. Consider the following true sentences: '7 + 5 = 12' and '7 > 3'. Both of these require the number 7 to exist, otherwise the sentences would be false. In standard semantics the objects denoted by singular terms in true sentences ('Mary', '7') exist. Consequently, mathematical objects do exist.[2]

(2) *Mathematical objects are outside of space and time*   The typical subject matter of natural science consists of physical objects in space and time. For pine trees, positrons and pussy-cats, we can always say *where and when*; not so for primes, π, or polynomials. The *standard metre* is kept in a special place in Paris; not so the number 27 which is to be found nowhere in space and time, though it is just as real as the Rock of Gibraltar. Some commentators like to say that numbers 'exist', but they don't 'subsist'. If this just means that they are not physical, but still perfectly real, fine. If it means something else, then it's probably just confused nonsense.[3]

(3) *Mathematical entities are abstract in one sense, but not in another*   The term 'abstract' has come to have two distinct meanings. The older sense pertains to universals and particulars. A universal, say *redness*, is *abstracted from* particular red apples, red blood, red socks, and so on; it is the *one* associated with the *many*. The notions of *group*, or *vector space* perhaps fit this pattern. Numbers, by contrast, are not abstract in this sense, since each of the integers is a unique individual, a particular, not a universal.[4]

On the other hand, in more current usage 'abstract' simply means outside space and time, not concrete, not physical. In this sense all mathematical objects are abstract. A simple argument makes this clear: There are infinitely many numbers, but only a finite number of physical entities; so most mathematical entities must be non-physical. It would seem rather unlikely that, say, the first n numbers are physical while from $n + 1$ on they are abstract. So, the reasonable conclusion is that all numbers, and indeed all mathematical objects, are abstract.[5]

(4) *We can intuit mathematical objects and grasp mathematical truths*
Mathematical entities can be 'seen' or 'grasped' with 'the mind's eye'. These terms are, of course, metaphors, but I'm not sure we can do better. The main idea is that we have a kind of access to the mathematical realm that is something like our perceptual access to the physical realm. This doesn't mean that we have direct access to everything; the mathematical realm may be like the physical where we see some things, such as white streaks in bubble chambers, but we don't see others, such as positrons.

This provides another great advantage of Platonism over some of its rivals, especially over conventionalist accounts. It explains the psychological fact that people feel the compulsion to believe that, say, $5 + 7 = 12$. It's like the compulsion to believe that grass is green. In each case we see the relevant objects. Conventionalists make mathematics out to be like a game in which we could play with different rules. Yet '$5 + 7 = 12$' has a completely different feel from 'Bishops move diagonally.' Platonism does much justice to these psychological facts.

(5) *Mathematics is a priori, not empirical*   Empirical knowledge is based (largely, if not exclusively) on sensory experience, that is, based on input from the usual physical senses: seeing, hearing, tasting, smelling, touching. Seeing with the mind's eye is *not* included on this list. It is a kind of experience that is independent of the physical senses and, to that extent, a priori.

There are profoundly different ways of being a priori. Conventionalist, formalist and intuitionist accounts of mathematics are all a priori, so this property does not differentiate them from Platonism until we look at the details – something I will do below shortly. Naturalism, however, is an exception; it's quite anti-a priori. Some recent accounts (Quine, Kitcher, and in some respects Maddy) want to assimilate mathematics to the natural sciences. One relevant fact in assessing this is that in the history of the various sciences, there has been much interaction among various branches. A revolution in physics can cause a subsequent revolution in chemistry or biology. Yet never in the entire history of mathematics has a result elsewhere (in the non-mathematical sciences) had any impact on our evaluation of mathematics. That is, no mathematical belief was ever overthrown by a discovery made in the natural sciences. I distinguish a logical conflict from suggestive interaction. Clearly, there has been a long and fruitful history of mutual stimulation. The discovery of non-Euclidean geometry, for example, did not entail any particular physics, though it did allow physicists to entertain a possibility that hadn't previously been considered. Notice the profound difference between that and the impact that quantum mechanics had on pre-quantum chemistry. (The mathematics–science relation is the subject of Chapter 4.)

(6) *Even though mathematics is a priori, it need not be certain*   These are quite distinct concepts. The mind's eye is subject to illusions and the vicissitudes of concept formation just as the empirical senses are. And axioms are often conjectures, not self-evident truths, proposed to capture what is intuitively grasped. Conjecturing in mathematics is just as fallible as it is elsewhere.

Not all versions of Platonism embrace fallibilism. Self-evidence and certainty have often been emphasized in the past. I'll take up this theme later in this chapter and show that the case for fallibility is quite favourable.

(7) *Platonism, more than any other account of mathematics, is open to the possibility of an endless variety of investigative techniques*   Proving theorems in a traditional way is certainly one method of establishing new mathematical truths, but it needn't be the only way, and Platonism does not stress it. In the natural sciences one might start from, say, the first principles of quantum mechanics and derive a new result. In this fashion we would come to know something new. But, of course, there are numerous other ways of learning new things about the physical world, including: direct observation, hypothesizing and testing the observable consequences, analogical reasoning, thought experimenting, and so on. Platonism can be similarly liberating for mathematical research.

There is no consideration, on a Platonistic account, that might lead us to doubt the existence and effectiveness of non-standard means for learning about the mathematical realm. But what other means might there be? What about conjectures that 'explain' several already known results? What about generalizations drawn from a large number of computer-generated instances? And especially, what about diagrams and pictures? Besides traditional proofs, these non-standard techniques may also bear much fruit. At the very least, these deserve a hearing. One of the chief aims of this volume is to explore such non-standard ways of investigating the mathematical realm.

Platonism does great justice to the mathematical image, as sketched in the introductory chapter. Not perfect justice, though. As I've outlined it, Platonism's fallibility is at odds with the traditional conception of mathematical certainty; and though it does not reject proofs, Platonism may not always require them. These differences may turn out to be big departures from the common view. In spite of these tensions, however, Platonism is remarkably close to the standard picture. And as we consider various accounts of the nature of mathematics, it will be apparent that Platonism is indeed much closer to traditional and common-sense views of mathematics than any of its rivals. Therefore it's no wonder that it has had an illustrious history and that working mathematicians are naturally sympathetic.

However, there are serious misgivings on the part of many with this view of mathematics. All would be happy in Plato's heaven were it not for a small cluster of problems connected with this question: If mathematical objects exist independent from us, outside space and time, then how could they be in any way accessible? and how could we come to know anything about them? I'll take up additional problems as we go along, but this is the big one.

The Problem of Access

The problem of access to the Platonic realm can be expressed in several different ways. Let's look at a couple of versions of this objection.[6]

> *We have a good understanding of the mechanism of normal perception, but no idea at all of the workings of 'the mind's eye', so postulating this sort of perception is dubious, perhaps even nonsensical.*

When it comes to ordinary perception, what exactly is understood? I see a coffee cup on my table. Photons come from the cup, enter my eye, interact with the rods and cones inside; a signal is send down the optical nerve into the visual cortex, and so on. This much – the physiological part of the process – is understood very well. But what about my *sensation* of the cup and my *belief* that there is a cup on the table? No one has the foggiest idea how these sensations and beliefs are formed. This is the mind–body problem – and it's utterly unsolved. There are lots of proposals – some ingenious and promising – but no one should be tempted to say that the mental part in this process (unlike the physiological part) is well understood. How the physical process brings about the belief is a very great mystery. It is just as great a mystery as how mathematical entities bring about mathematical beliefs. Of course, it would be wonderful to understand both, but our ignorance in the mathematical case is no worse than our ignorance in the case of everyday objects.

As an objection to Platonism, the claim that physical perception is well understood – even if completely right – seems off the mark. What, after all, is the Platonist claim? Crudely, this: We have mathematical knowledge and we need to explain it; the best explanation is that there are mathematical objects and that we can 'see' them. This is parallel to the following argument which is widely urged, for example, against Berkeley: We know the cup is on the table and we need to explain how this knowledge arose; the best explanation is that there really is a cup there and that we can see it – the cup causes our perception. Now consider: Do we need a detailed theory of perception for this argument to be persuasive? People rationally accepted this type of argument against Berkeley long before photons were discovered. The fact that Platonists have nothing to offer in the way of an explicit mechanism for seeing with the mind's eye does not in the least undermine the cogency of the Platonist claim. I hope no one was expecting an explanation in terms of little *platons* entering the mind's eye.

Let's turn to a related, but somewhat more specific objection.

*To know anything at all, there must be some sort of causal connection
between the object known and the knower. Abstract objects, being outside of space and time, are causally inert; so, we cannot possibly interact with them. Consequently, even if they exist, we could not know
them. Since Platonism is committed to our knowing abstract entities, it
is hopelessly wrong.*

In recent years the causal theory of knowledge has been popular with episte-
mologists, especially naturalists. The basic idea is very plausible. When we
consider how we come to know various things, we invariably seem to find some
sort of causal chain between the objects of knowledge and ourselves. If I know
that Mary is wearing a red shirt, it is because I am in causal contact: photons
from Mary enter my eyes, and so on. Even my knowledge of the past and future
is causally grounded. I know Brutus stabbed Caesar, because those in direct
causal contact recorded their experiences, and a chain of further recordings
leads to the printed page before me; then photons from that page enter my eye,
and so on. Moreover, even the future is in causal contact. I know, for example,
that it will snow tomorrow. The cloud formation I see now will shortly cause
snow. In this indirect way, I am causally connected to tomorrow's event. In rou-
tine cases – past, present or future – we seem able to pick out a causal chain
between ourselves and the object or event that we know about. What's more, if
this causal connection did not exist, it seems safe to say that we simply would
not have the knowledge in question.

These simple considerations make the causal theory of knowledge appear very
plausible. And when applied to abstract objects, it would seem that any causal
connection is completely lacking; so, according to this naturalistic account, we
cannot actually know anything about abstract entities after all, even if they do
exist. Our mathematical knowledge cannot be about abstract objects, but must
be understood in some quite different, non-Platonistic, way. This argument has
been used repeatedly in recent years[7] and it would seem that Platonists can only
retreat, reflect and be wretched.

But no, the argument is flawed. The best way to show that the causal theory of
knowledge is wrong is to show that – surprisingly – there is a case *within the
physical realm* where it fails miserably. If we can show that we have knowledge of some physical event, but that we cannot be in causal contact with that
event, then we will have shown that a causal connection is *not necessary* for
knowledge.

The case I have in mind comes from one of the bizarre situations which arise
in quantum mechanics. In an EPR-type set-up (from the thought experiment of
Einstein, Podolsky, and Rosen) a decay process gives rise to two photons moving in opposite directions towards detectors at either end of a room (Figure 2.1).
The detectors include polaroid filters which can determine whether the incom-
ing photons have the so-called property *spin-up* or *spin-down*. Both theory and

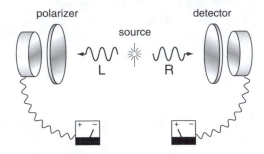

Figure 2.1 An EPR-type set-up

experience tell us that the two outcomes are always correlated: one photon is up and the other down. But we can never determine in advance which will arrive at either side – we seem to have complete randomness.

The interesting part is the perfect correlation, one up, the other down. Why does it occur? One possibility is that the measurement on one wing of the measuring apparatus *causes* the outcome at the other. However, we can rule this out by assuming with special relativity that no causal influences travel faster than light. The two measurements (made simultaneously in the frame of the laboratory) are outside of each other's light cones; so neither measurement causally affects the other. The second possibility is to assume that there must be something at the origin (at the time of the creation of the pair of photons) that is responsible for their correlated properties. (This was the EPR conclusion; such a common cause is known as a hidden variable.) Unfortunately, this common cause turns out to be impossible, too. The so-called Bell results show that such a common cause (i.e. a so-called local hidden variable) predicts a different measurement outcome than either quantum mechanics predicts or experience determines.[8]

The implications for the causal theory of knowledge are straightforward – it's false. Suppose I am at one wing of the measuring apparatus and get the result: spin-up. Then I can immediately infer that you at the other wing have the result: spin-down. I know the distant outcome *without being causally connected* to the remote wing. A direct causal connection would have to be faster than light, something ruled out by special relativity, and a common cause in the past grounding my knowledge would amount to a local hidden variable, something ruled out by the Bell results.

So the causal theory is simply refuted by this example. We can have knowledge even *without a causal connection*.

Of course, arguments like this are problematic. It relies on significant assumptions about the physical world. Perhaps special relativity is false; maybe some signals do go faster than light. This would make a direct causal connection possible after all. Perhaps the complex analysis of the situation involved in ruling out local hidden variables is flawed, making a common cause possible after all. These are indeed possibilities, but the assumptions that go into the physics of this situation are at least as plausible, or even more plausible, than the assumptions involved in the causal theory of knowledge. It seems perfectly sensible to dump the latter.[9]

Once the causal theory is rejected, there is no objection to our knowing about abstract entities without being causally related to them. The problem of access is a pseudo-problem; resistence to Platonism is motivated by misplaced scruples.

The Problem of Certainty

Philip Kitcher (1983) rejects Platonism for a number of reasons, one of which is the impossibility, as he sees it, of a priori knowledge.[10] His argument is simple: Consider a very long proof, so long that it would take a skilled mathematician months to work through it. Is there not some reasonable chance of an error? Of course there is. Consequently, one cannot be *certain* that the proof is right; the theorem is thrown into some measure of doubt. Thus, it cannot be known a priori.

Of course, this shows at most that some theorems (those with very long proofs) cannot be known a priori, but (as Kitcher allows), this leaves the rest untouched. So Platonism might still be a correct partial account. But even this is conceding too much to Kitcher. His argument only works by making the assumption that being a priori means being certain. There is no reason in the world to make that assumption – even though it has often been made in the past.

There are, of course, accounts of the a priori which lead rather naturally to its identification with certainty, so Kitcher is in good company. Formalists and positivists held that the source of a priori knowledge is to be found in language or in convention. 'Bishops move diagonally' and 'Bachelors are unmarried males' are true because *we stipulate* them to be so. That's how we play chess and how we speak English. Truths such as these are known a priori because they are based on decisions, not discoveries. And they are completely certain for the same reason – our saying makes them so. To this extent Kitcher is right: mathematics is not a priori in this sense. But this is not the only view of the a priori.

A Platonic realist holds that mathematics is not stipulative, but descriptive. A priori knowledge of mathematics then is knowledge which is not based essentially on empirical evidence. Seeing with the mind's eye is perhaps a kind of experience, but it is definitely not *sense* experience. So, in this way, it is a priori, not empirical. On the other hand, seeing with the mind's eye is fallible, just as normal sense experience is; so it does not result in knowledge which is certain. Fallible and a priori are perfectly compatible notions.

Another source of fallibilism is that mentioned by Gödel. He takes axioms to be conjectures, tested by means of intuitive consequences. In this regard it is

similar to, say, physics. We conjecture a theory, then draw out some consequences that can be checked in the lab. If the consequences turn out to be true,
our faith in the theory is increased; if the consequences are false, we consider
the theory refuted. Of course, this is extremely simplistic – testing is a subtle
business. But the main point holds: A false theory can have true consequences
as well as false ones, and we might be unlucky enough not to notice.

The history of set theory powerfully illustrates this point. In early versions
(now often called 'naive set theory'), the axiom of comprehension assumed that
for any condition there is a set of objects satisfying that condition. This means,
symbolically, $P(x) \leftrightarrow x \in \{x: P(x)\}$. This axiom seemed extremely plausible: 'x
is red' is equivalent to 'x is a member of the set of red things'. Alas, this principle leads directly to Russell's paradox when we let the condition be *is not a
member of itself*, i.e. $P(x) = x \notin x$.

The set of abstract objects A is itself an abstract object, so $A \in A$. The set B
of bananas is not itself a banana, so $B \notin B$. Some sets are members of themselves and some are not. So far, so good. Now let's form a set of all sets like B
that are not members of themselves. We'll call it R in honour of Russell and
define it as follows: $R = \{x: x \notin x\}$. Clearly, $B \in R$ and $A \notin R$. What about R?
Is it a member of itself or not?

When we ask does $R \in R$?, we find that if we assume Yes, $R \in R$, then we
have $R \in \{x: x \notin x\}$; therefore, $R \notin R$. And if we assume No, $R \notin R$, then we
have $R \notin \{x: x \notin x\}$; therefore $R \in R$. A contradiction. The way this is dealt
with in current set theory is to say the condition can only be defined on already
existing sets: *For any set A and condition defined on the members of A, there is
a set of objects satisfying that condition.* So the Russell set can no longer be
defined. The closest we can come is $R = \{x: x \in A \ \& \ x \notin x\}$, but this won't
lead to a paradox (at least, not as far as we know).

Imre Lakatos (1976) cheerfully combines fallibilism and Platonism (of a
sort). Lakatos is famous for his vigorous defence of 'empiricism' in mathematics; but this 'empiricism' is really just fallibilism. Nowhere does Lakatos say
mathematics is based on sensory input. Above, I gave a long list of ingredients
of Platonism similar to, but not identical with, the standard view of mathematics. Lakatos, I suspect, would accept all of these Platonistic ingredients. His
view is quite compatible with a fallibilistic version of Platonism and, indeed,
provides some of the best arguments for characterizing mathematics that way.

Lakatos focuses on the notion of a proof, which is the central device in *the
method of proofs and refutations*, as he calls it. It is at once a method both
descriptive and *prescriptive*, an account of the best mathematics of the past and
a guide to future research. The method is Lakatos's general heuristic guide to
mathematical discovery. In his eyes, proofs have more important things to do
than to justify theorems.

In his account, we start with a primitive conjecture, such as, that
$V - E + F = 2$ holds for any polyhedron (i.e. the number of vertices, V, minus

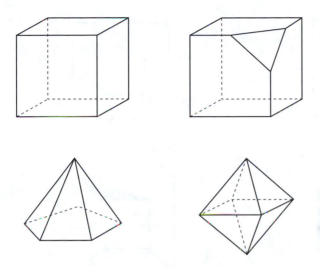

Figure 2.2 Various polyhedra; in each case V − E + F = 2

the number of edges, E, plus the number of faces, F, equals 2). The conjecture stems, perhaps, from noticing a pattern in several examples (Figure 2.2).

Next, a proof is given, and this is usually some sort of argument or, as Lakatos likes to call it, a thought-experiment. Often the proof takes the form of breaking things up into lemmas or subconjectures (which are tentatively accepted), and then of showing that these lemmas imply the initial conjecture.

TEACHER: I have [a proof]. It consists of the following thought-experiment. *Step 1:* Let us imagine the polyhedron to be hollow, with a surface made of thin rubber. If we cut out one of the faces we can stretch the remaining surface flat on the blackboard, without tearing it. The faces and edges will be deformed, the edges may become curved, but V and E will not alter, so that if and only if V − E + F = 2 for the original polyhedron, V − E + F = 1 for this flat network – remember that we have removed one face. [Figure 2.3 shows the flat network for the case of the cube.] *Step 2:* We now triangulate our map – it does indeed look like a geographical map. We draw (possibly curvilinear) diagonals in those (possibly curvilinear) polygons which are not already (possibly curvilinear) triangles. By drawing each diagonal we increase both E and F by 1, so that the total V − E + F will not be altered [Figure 2.4]. *Step 3:* From the triangulated network we now remove the triangles one by one. To remove a triangle we either remove an edge – upon which one face and one edge disappear [Figure 2.5(a)], or we remove two edges and a vertex – upon which one face, two edges and one vertex disappear (Figure 2.5(b)]. Thus if V − E + F = 1 before a triangle is removed, it remains so after the triangle is removed. At the end of this

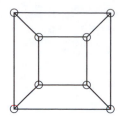

Figure 2.3

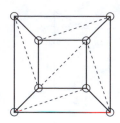

Figure 2.4

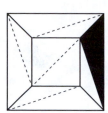

Figure 2.5(a)

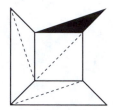

Figure 2.5(b)

procedure we get a single triangle. For this V − E + F = 1 holds true. Thus we have proved our conjecture.

DELTA: You should call it a *theorem*. There is nothing conjectural about it any more.

ALPHA: I wonder. I see that this experiment can be performed for a cube or for a tetrahedron, but how am I to know that it can be performed for *any* polyhedron? For instance, are you sure, Sir, that *any polyhedron, after having a face removed, can be stretched flat on the blackboard*? I am dubious about your first step.

BETA: Are you sure that *in triangulating the map one will always get a new face for any new edge*? I am dubious about your second step.

GAMMA: Are you sure that *there are only two alternatives – the disappearance of one edge or else of two edges and a vertex – when one drops the triangles one by one*? Are you sure that *one is left with a single triangle at the end of this process*? I am dubious about your third step.

TEACHER: Of course I am not sure.

ALPHA: But then we are worse off than before! Instead of one conjecture we now have at least three! You call this a 'proof'!

(Lakatos 1976: 7f)

Lakatos requires the mathematician to do two seemingly contradictory things: to prove the conjecture *and* to refute it. This does not mean to look for a proof, and failing that to seek a counter-example. Rather it means *to give a proof and to give a counter-example, as well*. The reason this apparently absurd dictum can actually be carried out is that, according to Lakatos, proofs do not incontrovertibly

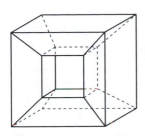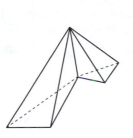

Figure 2.6 Some counter-examples to V − E + F = 2

prove, nor do counter-examples absolutely refute. (Figure 2.6 gives some counter-examples to the theorem. Readers can check this for themselves by simply counting the edges, faces and vertices and seeing if they satisfy V − E + F = 2. These examples will be discussed again in a later chapter.)

If proofs don't prove, and counter-examples don't refute, then just what do they do? Usually philosophers have been concerned with how theories come to be rationally accepted. On the other hand, how theories were thought of initially, is a question relegated to psychology. This is Reichenbach's (1938) widely known and accepted distinction between the 'context of discovery' and 'context of justification'. The latter can be successfully analysed by philosophers, but the former, claims Reichenbach, admits of no rational discussion – there can't be a rational method of having good ideas. But the method of proofs and refutations flies in the face of this. Not only do proofs justify (fallibly, of course), but they are the principle device in the generation of new concepts and theorems. In this fashion, discovery and justification become intimately connected. 'Proofs, even though they may not *prove*, certainly do help to *improve* our conjecture. . . . *Our method improves by proving. This intrinsic unity between the 'logic of discovery' and the 'logic of justification' is the most important aspect of the method*' (Lakatos 1976: 37).

In a nutshell, proofs are tools for concept formation. This is the key to understanding the kind of fallibilism that Lakatos is concerned with. He is not in the least worried about making computational mistakes, but rather with the fact that we are likely working with concepts of polyhedron, continuity, tangent space, fibre bundle, differential form, limit ordinal, and so on which are not quite right and will never be known to be right. Our concepts of electron, gene and subconscious have evolved considerably over the years. Mathematical concepts are similarly revisable, and therein lies a major source of fallibility in mathematics – perhaps the most interesting.

Russell's paradox, as I noted above, brought about a revision of an axiom (or at least implicit axiom; the paradox was discovered before set theory was officially axiomatized). But it did more than that. It also led to the revision of the concept of a set. Cantor famously said that a set is 'any collection into a whole M of definite, distinct objects m ... of our intuition or thought' (1895: 85). This clearly won't do as a conception of set, since it leads to the Russell set and on to the paradoxes. Currently, the reigning idea is the 'iterative conception of set'. We start with the empty set, φ (or perhaps with individuals which are not sets), then use various set-forming operations (characterized by the axioms) to build up ever more complex sets. These operations are legitimate only when applied to already existing sets. The thought 'set of all things which are not members of themselves', while legitimate on Cantor's conception of a set, is ruled out on the iterative conception. Contemporary set theorists are still up in the air over the proper conception of a set.[11]

This is a clear case of conceptual change. The fallibilism involved is that of having the wrong concept, not the fallibilism of having the wrong beliefs about the right concept.

I've sketched three sources of fallibilism in mathematics. There are mistakes stemming from the incorrect application of accepted principles (e.g. calculation errors). There are possibly false conjectures which have so far not been detected (e.g. our current set theory axioms might simply be false just as earlier versions were). And finally, there is the possible use of wrong or naive concepts (e.g. early concepts of polyhedron and set). Does any of this hurt Platonism? Not at all. Fallibilism ties in naturally with any sort of mathematical realism. Traditionally, Platonism has been linked to the epistemology of certainty. But the view of Platonism I've sketched here involves a methodology which is a lot more like the methodology of the natural sciences, one that involves a kind of perception – empirical observation in natural science, intuition in mathematics. And in each case it involves conjecture tested by observable or intuitive consequences, conceptual revision, and much more. How much more we'll see in the following chapters.

Platonism and its Rivals

No theory can be evaluated by a simple comparison with reality. Among other things, testing involves a comparison with rivals. So when thinking about Platonism, we must have an eye on formalism, constructivism, naturalism, and so on. This seems to be a general truth about intellectual life, but the situation in philosophy of mathematics is even more complicated. Some might concede that Platonism does an excellent job of accounting for everything in mathematics, yet still reject it. Why? It's because they hold a general philosophical outlook in which there is no room for Platonism. For example, a naturalist tries to account for things exclusively in terms of the *natural* sciences. This tends to be a reductionistic programme in which minds, morals, and mathematics are all to be explained, reduced to, or somehow understood in terms of the methods and ontology of the natural sciences. A naturalist might admit that Platonism is a better account of mathematics than naturalism when we confine our attention *narrowly* to the mathematical realm, but that naturalism, nevertheless, is to be preferred because it is *globally* a better account of things. When it comes to mathematics, then, a naturalist would claim to need only a *workable* account, not the best.

I have much sympathy with this general line of reasoning – not with naturalism, but with the idea that the total picture counts. Of course, I can't meet rivals head-on at a global level; that would require a book very much longer than this one which gives an account of how physics works, how linguistics works, how ethics works, and so on. But one should keep in mind that there are Platonistic accounts in these and other areas. For example, Platonistic accounts of ethics, of linguistics, of laws of nature, of thought experiments are all fairly promising in their own right.[12] Mathematics has always been Platonism's strong suit; but, as a general outlook, Platonism, I'm glad to say, is not faring badly at all.

Further Reading

Though Platonists have come a long way from Plato, it's always good to read the greatest philosopher who ever lived. The *Meno* and the *Republic* (Book IV) are the main sources for his views on mathematics. The leading Platonist of modern times is Gödel. His essays, 'Russell's Mathematical Logic' and 'What is Cantor's Continuum Hypothesis', are central.

Many books on the philosophy of mathematics have a chapter on Platonism, often critical. Here are a few that are worth reading and not just for their discussions of Platonism: Balaguer, *Platonism and Anti-Platonism in Mathematics*; Colyvan, *The Indispensibility of Mathematics*; Dummett, *Frege's Philosophy of Mathematics*; Frege, *Foundations of Arithmetic*; Hale, *Abstract Objects*; Irvine (ed.), *Physicalism in Mathematics*; Maddy, *Realism in Mathematics*; Potter, *Reason's Nearest Kin*; Russell, *Introduction to Mathematical Philosophy*; Resnik, *Mathematics as a Science of Patterns*; Shaprio, *Philosophy of Mathematics*.

This is but a handful from a much larger group of excellent books. Articles in the philosophy of mathematics can be found in many journals, especially those devoted to the philosophy of science. There is one excellent journal wholly devoted to the topic, *Philosophia Mathematica*.

CHAPTER 3
Picture-proofs and Platonism

Mathematicians, like the rest of us, cherish clever ideas; in particular they delight in an ingenious picture. But this appreciation does not overwhelm a prevailing scepticism. After all, a diagram is – at best – just a special case and so can't establish a general theorem. Even worse, it can be downright misleading. Though not universal, the prevailing attitude is that pictures are really no more than heuristic devices; they are psychologically suggestive and pedagogically important – but they *prove* nothing. I want to oppose this view and to make a case for pictures having a legitimate role to play as evidence and justification – a role well beyond the heuristic. In short, pictures can prove theorems.[1]

Bolzano's 'Purely Analytic Proof'

Bernard Bolzano proved the intermediate value theorem. This was early in the nineteenth century, and commentators since typically say two things: first, that Bolzano's work was initially unappreciated and only later brought to light or rediscovered by others such as Cauchy and Weierstrass; second, that thanks to Bolzano and the others, we now have a *rigorous proof* of the theorem, whereas before we only had a good hunch based on a geometrical diagram.

Typical advocates of this view are the historians Boyer (1949) and Kline (1972) who, respectively, discuss Bolzano in chapters called: 'The Rigorous Formulation' and 'The Instillation of Rigor'. It's easy to guess from these titles where their hearts lie and how appreciative their view of Bolzano's efforts might be. Mathematicians hold a similar outlook. Most calculus and analysis texts contain a proof of the intermediate value theorem, and often they have a few casual comments about its significance. Apostol, for example, remarks: 'Bolzano . . . was one of the first to recognize that many "obvious" statements about continuous functions require proof' (1967: 143). Courant and Robbins, in praising Bolzano, say 'Here for the first time it was recognized that many apparently obvious statements concerning continuous functions can and must be proved if they are to be used in full generality' (1941: 312).

The common attitude towards Bolzano reflects the generally accepted attitude towards proofs and pictures. On this view only proofs give us mathematical knowledge; moreover, proofs are derivations; they are verbal/symbolic entities. Pictures, on the other hand, are psychologically useful, often suggestive, and sometimes downright charming – but they do *not* provide evidence. When this attitude is brought to bear on the intermediate value theorem, it's perfectly natural to conclude that – until Bolzano – we couldn't really be sure the theorem was true.

Let's look at one of three related theorems (sometimes called the intermediate zero theorem) due to Bolzano (1817).

> *Theorem*: If f is continuous on the interval $[a,b]$ and f changes sign from negative to positive (or vice versa), then there is a c between a and b such that $f(c) = 0$.

Here is a proof which, while not exactly Bolzano's, is in the modern spirit which he created.[2]

> *Proof*: Assume (with no loss of generality) that $f(a) < 0 < f(b)$. Let $S = \{x: a \leq x \leq b \ \& \ f(x) < 0\}$. This set is not empty, since a is in it; and it is bounded above by b, so it has a least upper bound, c. There are three possibilities.

> (1) $f(c) < 0$. If this is true there is an open interval around c, i.e. $(c - \delta, c + \delta)$, in which $f(x) < 0$, for all x in the interval including those greater than c. This contradicts the assumption that c is an upper bound.
> (2) $f(c) > 0$. If this is true there is an open interval around c, i.e. $(c - \delta, c + \delta)$, in which $f(x) > 0$, for all x in the interval, even those less than c. But that's impossible since c is the least of all the upper bounds, so that $f(x) < 0$ for all x less than c.
> (3) $f(c) = 0$. The other two possibilities being ruled out, this one remains. And so, the theorem is proved.

Consider now visual evidence for the theorem. Just look at the picture (Figure 3.1). We have a continuous line running from below to above the x-axis. Clearly, it *must cross* that axis in doing so. Thus understood, it is indeed a 'trivial' and 'obvious' truth.

A simple generalization of this theorem leads to what is now known as the intermediate value theorem, also proved by Bolzano.

Theorem: If *f* is continuous on the interval [*a*,*b*] and there is a *C* between *f(a)* and *f(b)*, then there is a *c* between *a* and *b* such that *f(c)* = *C*.

I won't bother to give a proof in the Bolzano style, but I will provide another picture (Figure 3.2).

Bolzano also gives a third theorem, again, a generalization from the others.

Theorem: If *f* and *g* are both continuous on the interval [*a*,*b*] and *f(a)* < *g(a)* and *f(b)* > *g(b)*, then there is a *c* between *a* and *b* such that *f(c)* = *g(c)*.

Once we have the hang of the first theorem, we can easily extend the result to the second and third using the same techniques. Again, I'll forgo the analytic proof, but not the visualization. However, this time I'll call on your imagination; like Shakespeare's Prologue on a stage serving as the imagined battlefield of Agincourt, I'll urge you to 'Work your thoughts!' Consider this little problem: A mountain climber starts at the base of a mountain at noon and reaches the top at 6 p.m. She sleeps the night there, then at noon the next day, returns to the bottom following the same path. Question: Is there a time at which she was at the same point on the mountain path both days? The answer, surprisingly, is Yes, in spite

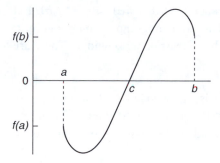

Figure 3.1 The intermediate zero theorem

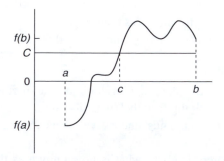

Figure 3.2 The intermediate value theorem

of how fast she may go up or down the hill. Here is how to solve the problem: Consider an equivalent situation in which we have two hikers, one at the top heading down, the other at the bottom heading up, both setting out at noon on the same day. Obviously, they eventually meet somewhere on the path. And when they do, that is the common time. The solution to the riddle perfectly illustrates the third theorem. It also proves it. Bolzano, of course, gave a 'purely analytic proof', as he called it, not a visualization.

What Did Bolzano Do?

There is a spectrum of ways to understand Bolzano's achievement. The first of these is the common view I mentioned above.

(1) *Bolzano firmly established a theorem that was not known to be true until his proof.* The diagram, on this account, perhaps played an important heuristic role, but nothing more. Not only is this a common view of the matter, but some of Bolzano's own remarks about the fallibility of geometric intuition strongly suggest that this is how he viewed things.[3] But, of course, this is absurd. The geometric picture gives us a very powerful reason for believing the result quite independently of the analytic proof. Using the picture alone, we can be certain of this result – if we can be certain of anything.

Quite aside from the virtues or vices of pictures, we ought to have a some-what more humble attitude towards our understanding of verbal/symbolic reasoning. First-order logic may be well understood, but what passes for acceptable proof in mathematics includes much more than that. Higher-order logic is commonplace, but is far from being house-broken. Moreover, proofs are almost never given in full; they are just sketches which give ample scope for committing some of the well-known informal fallacies. Pictures can sometimes even expose verbal fallacies. As for Bolzano in particular, the principles that he used included naive set theory, now known to be profoundly inconsistent. A dose of humility seems called for.

Consider now a second view of what was achieved.

(2) *Bolzano's proof explained the theorem.* Imre Lakatos (1976) often talks this way about mathematics in general, and would, perhaps, endorse such a view of Bolzano. Philip Kitcher (1975, 1983) holds it explicitly and to some extent so does Alberto Coffa (1991). The verbal/symbolic proof may well explain the theorem, but the picture explains it, too – at least it explains why the continuous function cuts the x-axis (i.e. at $y = 0$) somewhere or other. The fact that the analytic proof explains the theorem does not set it apart from the picture. (This is explained in more detail below.) Now a third possibility:

(3) *The theorem confirms Bolzano's proof.* Bolzano is generally considered the 'father of arithmetization', as Felix Klein called him. The arithmetization

programme of the nineteenth century sought to found all of analysis on the concepts of arithmetic and to eliminate geometrical notions entirely. (The logicism of Frege and Russell carried this a step further in trying to reduce arithmetical notions to logic.) Proving something *independently known to be true* was then a feather in the cap of this programme. This method has many champions. Gödel, for example, thought new axioms for set theory should be accepted on the basis of their 'fruitfulness', that is, their good consequences, not their self-evidence. Russell, too, expressed the view clearly: 'we tend to believe the premises because we can see that their consequences are true, instead of believing the consequences because we know the premises to be true' (1907: 273f.).

It is pretty clear, that of our three options, the final one is the best. (The second option, explanation, is compatible with the third, confirmation, but can't be the whole story.) The consequence of adopting (3) is highly significant for our view of pictures. We can draw the moral quickly: *Pictures are crucial*. They provide the *independently-known-to-be-true* consequences that we use for testing the hypothesis of arithmetization.[4] Trying to get along without them would be like trying to do theoretical physics without the benefit of experiments to test conjectures.

Different Theorems, Different Concepts?

A pair of objections to all this is possible. One objection is that we have two different proofs of the same result, each with its own strengths and weaknesses. This I think is quite true – but it is not really an objection. I could rephrase (3) as saying: the two proof techniques arrive at the same result. One of these (the picture) is prima facie reliable. The other (the analytic proof) is questionable, but our confidence in it as a technique is greatly enhanced by the fact that it agrees with the reliable method.

I should add that the way the picture works is much like a direct perception; it is not some sort of encoded argument. However, the boundary between these two ways of understanding the pictures may not be very sharp, since even in fairly simple direct perceptions some 'interpreting' goes on. Ultimately, it may not matter which way we construe the picture, so long as the encoded argument (if there is one) is not the same argument as that given by the verbal/symbolic proof. For, either way, the picture serves as independent evidence for Bolzano's arithmetization programme.

The second possible objection is that we actually have different concepts of continuity at work: one is the ε-δ concept, which is more or less Bolzano's; the other is so-called pencil continuity, a geometric notion. To some extent this point seems right; we do have different conceptions of continuity. However, it would be a mistake to infer that the results of the two proofs are *incommensurable*. For one thing, if they are totally unrelated concepts then it would make no sense even to illustrate Bolzano's theorems with the diagrams, nor would it make any sense to apply Bolzano's result to situations in geometry or mechanics, as is commonly done. If we did take the attitude that these are two quite distinct conceptions of continuity, then we would be very hard pressed to account for a significant amount of mathematical practice. Even if the picture merely does psychological work, that in itself could only be explicable by assuming that ε-δ continuity and pencil continuity are somehow deeply related. If they are completely unrelated, then what is the picture doing there? It would be like a dictionary giving a verbal description of apples but illustrating the definition with a picture of a banana.

Perhaps a better understanding of what has historically transpired would be similar to our understanding of what happens in physics or biology. Theories in the natural sciences are tested by observations; however, those very observations are theory-laden. In the act of theorizing about some phenomenon we transform the description of the phenomenon itself (ducks to rabbits). However, the phenomenon – now under a different description – is still relevant for testing. In the case of Bolzano, perhaps the same thing has happened. The concept of continuity has changed, but the diagram is still relevant for testing purposes.

Inductive Mathematics

It's an uncontentious fact that mathematical reasoning is broader than merely proving theorems. We sometimes forget this when emphasizing the great achievements of mathematics and the ingenious proofs that have established our most treasured results. But obviously there is more. After all, why work on *this* problem rather than *that* one? Why fund this line of research rather than some other? Mathematicians and the mathematical community make all sorts of decisions which are not based on solid analytic proofs. A certain line of research on the Riemann hypothesis is financially supported – not because it's known to be correct, but because it seems promising. Another is rejected as a dead end. Where do these judgements come from? What grounds them? What is the basis of these attitudes which are so crucial to mathematical activity?

Let us call any evidence which falls short of an actual traditional proof 'inductive evidence'. Mathematical *achievements* may rest entirely on deductive evidence, but mathematical *practice* is based squarely on the inductive kind. Let's now look briefly at some types. (I'll look at some of these again in later chapters.)

Enumerative induction   Goldbach's conjecture says that every even number (greater than 2) is the sum of two primes. Check some examples: $4 = 2 + 2$, $6 = 3 + 3$, $8 = 5 + 3$, $10 = 5 + 5$, $12 = 5 + 7$, and so on. Computers have been used to check this well into the billions. No counter-examples have been found so far. Mathematicians tend to believe that Goldbach's conjecture is true. They don't have a proof, but they do have strong inductive evidence.

Analogy   Euler found a way to sum an infinite series that is not 'rigorous' by any stretch of the imagination. He argued from analogy that

$$\sum_{n=1}^{\infty} \frac{1}{n^2} = \frac{\pi^2}{6}.$$

Polya (1954: 17ff.) celebrated Euler's accomplishment, and Putnam (1975) endorsed it, too. Euler's reasoning was ingenious and persuasive, but not a proof. (This example is discussed in more detail in Chapter 10.)

Broad experience   Pose a problem; attack it from every conceivable angle; if all plausible approaches lead nowhere, it's time to think the initial conjecture false. The question, Is it true that P = NP?, was first posed about thirty-five years ago. This is the central problem for those working on computational complexity. The issue concerns how fast computational problems grow with size of input. There is now a broad consensus in the field that P ≠ NP. Of course, there is no proof, but a grant proposal which hoped to produce a positive result would be turned down flat.[5]

These kinds of inductive considerations are central to mathematical activity. Of course, someone could cheerfully grant this sort of thing and then appeal to the traditional distinction between 'discovery' and 'justification'. Inductive evidence, one might claim, plays a role in thinking up theorems, but proofs (and only proofs) give us real justification. (The only difference between this and the distinction philosophers champion is that the distinction here allows the existence of a 'logic of discovery' which philosophers often deny.)

But when we turn back to the subject of proofs, we quickly encounter a problem. Analytic proofs, after all, aren't constructed *ex nihilo*; they are based on axioms or first principles. But where do these first principles come from? Why do we believe these axioms? Once 'self-evidence' was an acceptable answer, but no more. I mentioned Gödel and Russell above. The Russell passage deserves quoting in full:

> we tend to believe the premises because we can see that their consequences are true, instead of believing the consequences because we know the premises to be true. But the inferring of premises from consequences is the essence of induction; thus the method of investigating the principles of mathematics is really an inductive method, and is substantially the same as the method of discovering general laws in any other science.
>
> (Russell 1907: 273–74)

Gödel shares Russell's consequentialist outlook; that is to say, he, too, holds that first principles are believed because they have the right consequences, not because they themselves are evident.[6] But, of course, this view only works because at least *some* of the consequences are evident. Mathematical intuition, as it is often called, must play a role. There are some mathematical truths that are obvious. Gödel and Russell argue that arriving at first principles or axioms in mathematics is similar to science. Mathematical intuitions are like empirical observations in physics. A system of axioms, say for set theory, is postulated just as a theory, say quantum mechanics, is postulated in physics. The theory (in either case) is tested by deriving consequences from it, and is supported by consequences which are intuitive or observational truths, while intuitive or observational falsehoods would refute the theory.

The intuitive truths of mathematics need not be certainties any more than ordinary empirical observations must be incorrigible to be confidently used by scientists. The parallel postulate need not be embraced in spite of its intuitive character. And Russell's paradox (which was explained in the last chapter) shows us that some things which seem highly evident (i.e. that sets exactly correspond to properties) are, in fact, downright false. Still, we can use these intuitions, just as we can use our ordinary eyesight when doing physics, even though we sometimes suffer massive illusions.

The relation for Gödel between a general theory (such as the axioms of set theory) and individual intuitive truths is one of reflective equilibrium, to use a notion introduced by Goodman and made famous by Rawls. That is, we try to construct a theory which is maximally powerful, simple, etc. and which does maximal justice to the intuitive truths. But we allow the possibility that a great mathematical theory will overrule a mathematical intuition, just as a great scientific theory will sometimes overrule an experimental result. The axiom of choice, for example, is widely accepted today, in spite of some bizarre consequences such as the Tarski–Banach paradox.

Even though such famous logicians as Russell and Gödel advocate this view, it has been relatively uninvestigated. Just what is the relation between axioms and intuitions? Should we characterize it as simple H-D? Or perhaps Bayesian? Is Popper's conjectures and refutations model the right one? Should the intuitive truths be 'novel', or can they be already known? These questions have gone largely unexplored, though Lakatos (1976) is a notable exception.

However, it's not this, but something else in Gödel's account that I want to focus on, namely the 'perception' of mathematical truths. Observational evidence in physics tends to consist in singular space-time observations: 'This object, here-now, has property such and such.' Mathematical intuitions are similar; they are relatively concrete and tend to be singular rather than general (e.g. '5 + 7 = 12'), though this is certainly not invariable. One thing that pictures in particular might do is greatly enlarge the pool of intuitive truths and perhaps even vary their character by adding ones that are relatively more general.

Special and General Cases

We learn early on not to confuse the special case with the general. Yet there are remarkable examples where the special is equivalent to the general. (And not just in trivial cases where the domain has only a single member, thus forcing $\forall x Fx \leftrightarrow \exists x Fx$.) The following is an example stressed by Polya (1954). First, consider the Pythagorean theorem which says that the square on the hypotenuse is equal to the sum of the squares on the other two sides, $c^2 = a^2 + b^2$. Next consider a generalization of this, $rc^2 = ra^2 + rb^2$, which says (see Figure 3.3) that for similar figures described on a right-angled triangle, the area on the hypotenuse is equal to the sum of areas on the other two sides. Clearly, the Pythagorean theorem is a special case of the more general theorem, arrived at by letting $r = 1$. We can move back and forth from special case to general case with ease.

But now comes a more interesting example. The right-angled triangle is itself a special case of the more general theorem. We divide the triangle into two, making three similar right-angled triangles. Think of each as an area constructed on its own hypotenuse, but lying *inside* the main triangle. Clearly the area of the whole triangle equals the area of the other two. The special case leads to the general case which in turn implies the Pythagorean theorem. Thus, we have a proof that is at root quite obvious – a simple observation.

There are two things I want to get out of this example. One thing is just to provide another illustration of a picture-proof. But more important is a second feature, the equivalence of the special case with the general case. Polya remarks that the example shows something 'surprising to the beginner, or the philosopher who takes himself for advanced, that the general case can be logically equivalent to a special case' (1954: 17). We can overlook the slight to philosophy, but not the remark that they are 'logically equivalent'. They are not. Rather they are mathematically equivalent in the sense that given certain mathematical assumptions about the distribution of r over $+$ and about the geometry of similar figures, the various cases are equivalent. (Symbolically, instead of $\vdash S \leftrightarrow G$, we have $\Gamma \vdash S \leftrightarrow G$.) However, I do not want to blunt Polya's significant point. There is something very important about the equivalence of

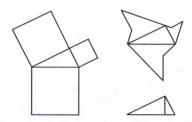

Figure 3.3 The special and the general cases are equivalent

the special and the general cases when they do occur, even if powerful assumptions are at work in the background. Often these background assumptions are not articulated until well *after* the equivalence has been established, perhaps they even arise only in order to explain the surprising nature of the equivalence. At any rate, an important principle of inference is at work here. And, of course, pictures play a crucial role in it.

Instructive Examples

I'll give some example theorems from number theory and infinite series – places where one would least expect to find instructive pictures. In each case the proof will be a diagram. The things to look for are these: Is the diagram convincing? Is it a special case (i.e. for some particular n)? And does it establish complete generality (i.e. for every n)? Would a standard verbal/symbolic proof of the theorem, say by mathematical induction, be more convincing?

Theorem: $1 + 3 + 5 + \ldots + (2n - 1) = n^2$

Proof:

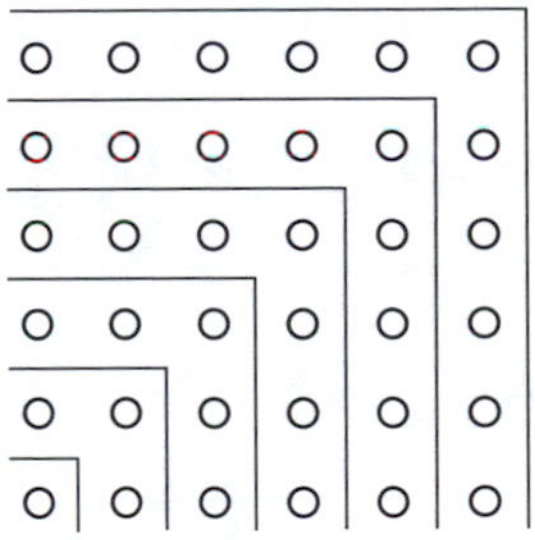

Figure 3.4

This picture-proof should be contrasted with a traditional proof by mathematical induction which would run as follows:

Proof (traditional): We must show first, that the formula of the theorem holds for 1 (the basis step), and second, that if it holds for n then it also holds for $n + 1$ (the inductive step).
Basis: $((2 \times 1) - 1) = 1 = 1^2$
Inductive: Suppose $1 + 3 + 5 + \ldots + (2n - 1) = n^2$ holds as far as n. Now we add the next term in the series, $2(n + 1) - 1$, to each side:

$$1 + 3 + 5 + \ldots + (2n - 1) + 2(n + 1) - 1 = n^2 + 2(n + 1) - 1$$

Simplifying the right-hand side, we get:

$$n^2 + 2(n + 1) - 1 = n^2 + 2n + 2 - 1$$
$$= n^2 + 2n + 1$$
$$= (n + 1)^2$$

This last term has exactly the form we want. And so the theorem is proven.

Theorem: $1 + 2 + 3 + \ldots + n = \dfrac{n^2}{2} + \dfrac{n}{2}$

Proof:

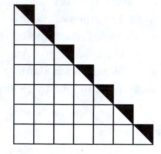

Figure 3.5

Again, for the sake of a contrast, here is a traditional proof.

Proof (traditional):

Basis: $1 = \dfrac{1^2}{2} + \dfrac{1}{2}$

Inductive:
$$1 + 2 + 3 + \ldots + n = \frac{n^2}{2} + \frac{n}{2}$$

$$\therefore 1 + 2 + 3 + \ldots + n + (n + 1) = \frac{n^2}{2} + \frac{n}{2} + (n + 1)$$

$$= \frac{n^2}{2} + \frac{n}{2} + \frac{2n}{2} + \frac{2}{2}$$

$$= \frac{n^2 + 2n + 1}{2} + \frac{n + 1}{2}$$

$$= \frac{(n + 1)^2}{2} + \frac{(n + 1)}{2}$$

Theorem: $\dfrac{1}{2} + \dfrac{1}{4} + \dfrac{1}{8} + \ldots = 1$

Proof:

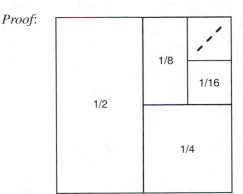

Figure 3.6

For the sake of a contrast, here is a standard proof using ε-δ techniques:

Proof (traditional): First we note that an infinite series converges to the sum S whenever the sequence of partial sums $\{s_n\}$ converges to S. In this case, the sequence of partial sums is:

$$s_1 = \frac{1}{2}$$

$$s_2 = \frac{1}{2} + \frac{1}{4}$$

$$s_3 = \frac{1}{2} + \frac{1}{4} + \frac{1}{8}$$

$$s_n = \frac{1}{2} + \frac{1}{4} + \frac{1}{8} + \ldots + \frac{1}{2^n}$$

The values of these partial sums are:

$$\frac{1}{2}, \frac{3}{4}, \frac{7}{8}, \ldots, \frac{2^n-1}{2^n}.$$

This infinite sequence has the limit 1, provided that for any number $\varepsilon > 0$, no matter how small, there is a number $N(\varepsilon)$, such that whenever $n > N$, the difference between the general term of the sequence $\frac{2^n-1}{2^n}$ and 1 is less than ε. Symbolically,

$$\lim_{n \to \infty} \frac{2^n-1}{2^n} = 1 \text{ iff } (\forall \varepsilon)(\exists N) n > N \to \left| \frac{2^n-1}{2^n} - 1 \right| < \varepsilon$$

A bit of algebra gives us the following:

$$\left| \frac{2^n-1}{2^n} - 1 \right| < \varepsilon \leftrightarrow \left| \frac{-1}{2^n} \right| < \varepsilon$$

$$\leftrightarrow 2^n > \frac{1}{\varepsilon}$$

$$\leftrightarrow \log_2 \frac{1}{\varepsilon} < n$$

Thus we may let $N(\varepsilon) = \log_2 \frac{1}{\varepsilon}$

Hence, $n > \log_2 \frac{1}{\varepsilon} \rightarrow \left| \frac{2^n - 1}{2^n} - 1 \right| < \varepsilon$

And so, we have proved that the sum of the series is 1.

The next example is particularly nice.

Theorem: $\sum\limits_{n=1}^{\infty} \frac{1}{4^n} = \frac{1}{4} + \frac{1}{16} + \frac{1}{64} + \frac{1}{256} + \ldots = \frac{1}{3}$

Proof:

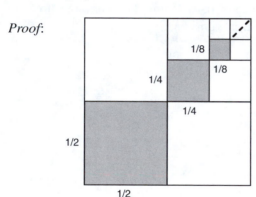

Figure 3.7

More examples of picture proofs are included in the Appendix to this chapter.

Representation

It's probably true that anything can stand for anything. But it's not true that anything can stand *pictorially* for anything. Something special is needed. But what is it about a picture of X that makes it a *picture* of X? The problem is related to the problem of intentionality in language and mind: How do words or thoughts get to be *about* things? How do they *represent*? Similarly, how do pictures represent the things they are pictures of?

There is a wide spectrum of views involved in a full answer to these questions, issues involving intentionality, conventions, and so on. I won't take up these issues but instead simply focus on a view of pictures which seems highly plausible, at lest initially.

In the *Tractatus*, Wittgenstein made a few cryptic remarks about the relation between pictures and what they picture. 'For a picture to work there must be something in common with what it pictures – "pictorial form"' (*Tractatus*, 2.161). 'The minimal commonality between pictorial form and object is *logical form*' (*Tractatus*, 2.18). What this suggests is a kind of structural similarity, a notion which is captured by the concept of an *isomorphism*. Barwise and Etchemendy (who are among the very few sympathetic to the use of pictures in inference) explicitly adopt such a view. They hold that 'a good diagram is isomorphic, or at least homomorphic, to the situation it represents' (1991: 22). Hammer (1995) explicitly adopts this account.

Let me take a moment to explain these notions. *Isomorphism* and *homomorphism* are usually defined for the particular case at hand, i.e. 'group-homomorphism', 'ring-isomorphism', or 'isomorphism of Boolean algebras', and so on. But there's a common idea involved. Two structures are *isomorphic* when (a) they have the same number of elements or objects, and (b) the relations among the elements of one structure have the same pattern as the relations among the elements of the other. Suppose three people, A, B and C are sitting around a table playing cards. A is to the left of B, B is to the left of C, and C is to the left of A. So the structure, call it S, consists of three objects and the relation 'is to the left of'. Now imagine another structure, S', with three elements, X, Y and Z, my family pets, a dog, a bird and a hamster, respectively. They terrify one another in the following order: the dog frightens the hamster, the hamster frightens the bird, and the bird frightens the dog. We set up the following correspondence: A $\leftrightarrow$ X, B $\leftrightarrow$ Y, C $\leftrightarrow$ Z, and we note that when A is to the left of B, then X frightens Y, and so on. Thus, we can conclude that S and S' are isomorphic. A *homomorphism* is a weaker notion. It requires the second condition, but not the first; homomorphic structures may have different numbers of elements. Now back to the proposal of Barwise and Etchemendy and of Hammer.

In a wide variety of cases, the proposal seems exactly right. Arguably, it holds, for instance, in the case of the two infinite series examples given above. But in general this is not so. Consider again the picture-proof of the number theory result which was given above. Notice, however, that it is just a picture (in the normal sense) of the $n = 7$ case; and so we can claim an isomorphism to some number structure with that cardinality. It is certainly not, however, isomorphic to all the numbers. True, it is homomorphic to the whole number structure. But note that a homomorphism to a larger structure is (at least in the case at hand) an isomorphism to a part. The picture (on the Barwise–Etchemendy and Hammer account) tells us about the isomorphic part, but sheds no light on the rest. For example, our picture (Figure 3.8(a)) is homomorphic to Figure 3.8(b). But we can't make inferences from the picture (3.8(a)) to the non-isomorphic part

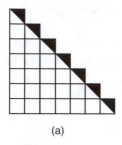

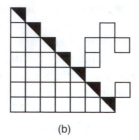

(a)   (b)

Figure 3.8

(in 3.8(b)) at all. There is, I claim, no useful homomorphism from our picture to *all* the natural numbers and no isomorphism at all. But still the diagram works. It does much more than establish the formula for $n = 7$; it establishes the result for all numbers.

Consequently, I want to suggest something quite different. My bold conjecture (to use Popper's terminology) is this: *Some 'pictures' are not really pictures, but rather are windows to Plato's heaven.* The number theory diagram is certainly a representation for the $n = 7$ case, but it is not for all generality. For the latter, it works in a different way, more like an instrument. This, of course, is a realist view of mathematics, but not a realist view of pictures. As telescopes help the unaided eye, so some diagrams are instruments (rather than representations) which help the unaided mind's eye.

Seeing Induction?

Let me quickly try to deal with a potential objection. The diagram (Figure 3.4 or 3.8(a)) that provided the proof of the theorem could be interpreted in a kind of Kantian way. The claim is this: one sees in the picture the possibility of a reiteration; the diagram can be extended to any number; that's why it works. The objection is anti-Platonistic in that it makes a Kantian point about constructability.

A curious fact about the diagram is that it strikes people in two distinct ways. Some see mathematical induction in it, others don't. I'm in the latter group. Whenever I have shown the example to philosophical or to mathematical audiences, there is a split into two roughly equal camps. Some mathematicians (and similarly some philosophers) are often quite adamant that induction *definitely is* or that it *definitely is not* encoded into the diagram. The common claim of some who see induction in the picture is that the picture is indeed legitimate as a proof, but it is so because of induction, not because of some Platonic reason. Others say that the picture is really a heuristic device that suggests mathematical induction, and that it is induction itself that is the genuine and legitimate proof of the theorem. These are interesting objections and I'll say three things in reply.

First, my view about pictures is two-fold: that they can play an essential role in proofs *and* that there is a Platonistic explanation for this. One version of the Kantian objection (or the it-is-really-just-induction objection) is only in opposition to the second, Platonistic, aspect. Pictures, and seeing the possibility of constructions, can still be a legitimate form of mathematical proof. Indeed, the legitimacy of pictures is upheld by the objection.

Second, the different interpretations of how the picture works are related to the distinction between potential and actual infinities. Both see the formula as holding for $\forall n \in \omega$. But the Kantian iteration account sees ω as a potential infinity only; the Platonistic account sees ω as an actual or completed infinity. Of course, the proper understanding of infinity is an unsettled question, but classical mathematics (especially set theory) seems committed to actual infinities. So I see my Platonistic interpretation of how the picture works as being favoured for that reason over the Kantian one.

Third, we might well wonder: What has the perceived possibility of constructing a diagram of any size got to do with numbers? I certainly don't deny that we can see the possibility of indefinite iterations of the diagram, but the Kantian objection seems to assume that we know the number theory result *because* we see the possibility of iteration. I don't know of any argument for this. We could just as well claim that we see the possibility of iteration *because* we have the prior perception of the number theory result.

Charles Chihara, takes up the induction theme (as he found it in the first edition of this book) and explains it as follows.

> Suppose that it is now said: 'Yes, I can see that the formula will tell one the number of squares in row n for each of the rows in the diagram (up to, that is, $n = 7$). But how do you know it will work no matter what n is?' Here let us consider how one could continue expanding the diagram to obtain the above result for $n = 8, 9, 10, \ldots$ To go from the nth case to the $n + 1$st case, one merely adds a row to the diagram. In particular, to get the $n + 1$st row, just copy the nth row directly below and then add one more square to the right end of the row (and darken the triangular area that lies to the upper right of the diagonal, as was done in the others). Thus, in order to obtain a diagram of the $n + 1$st case, one will just add a row consisting of exactly $n + 1$ squares. So the number of squares in row $n + 1$ will be $n + 1$. By an intuitive version of mathematical induction, we see that the formula for how many squares that are in row n works for all the natural numbers.
>
> Thus, to calculate the number $(1 + 2 + 3 + \ldots + n)$, one need only to calculate the total number of squares in the diagram which has n rows.

But this will be $n^2/2$ (the number of squares in the isosceles triangle whose side is n) plus $n/2$ (the number of squares which the darkened triangles in the diagram make up). The diagramatic proof is completely convincing.

(Chihara 2004, 302f)

Chihara concludes that there is no need to accept my Platonistic account of how the diagram works, since his inductive version is perfectly adequate. He also could have added that his way of accounting for the picture is close to how a standard proof would go, a bonus for his account. In short, there's much to be said for Chihara's account, if the picture proof works as he says it does.

For now, I will accept his claim (but only for the sake of the argument) that induction is somehow suggested in the diagram. Nevertheless, I will argue that a Platonistic account is still required.

The diagram is for the specific number $n = 7$. If we see induction in that diagram, it can only be that we know how to extend it to $n = 8$. So far, so good. But that only gives us a statement of the form: $S(7) \rightarrow S(7+1)$; i.e. if the picture proof holds for the number 7, then it holds for the number $7 + 1$. What possible grounds could one have for further claiming that if any diagram illustrated the case of an arbitrary number n, then that same diagram could be extended to the case of $n + 1$? I don't doubt that this is true, but one would be making a fantastic leap, since there are infinitely many n involved. It is certainly not something that is perceived with the senses.

We pass from a single finite case to a conclusion involving infinitely many numbers. How is the leap possible? We could try to explain it in terms of the earlier theory of pictures that I described above. That is, we could say the picture and its object have a similar structure and we could try to make this precise with the notions of isomorphism or homomorphism. But we shall run into the same problems as before. Cardinality considerations once again rule out an isomorphism. And, just as we saw previously, there are too many homomorphisms. Why is it that we seem to grasp the right infinite structure?

We might even liberalize the diagram a bit, using standard conventions involving dots, '. . .' (Figure 3.9). Still, the same considerations arise. We can see (in the literal physical sense) a finite case, but somehow end up seeing (in the 'mind's eye' sense) the infinite case.

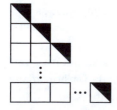

Figure 3.9

At this point, the only conclusion to draw seems to be this: If we see mathematical induction in the diagram, it is because the diagram is not actually a pictorial representation of induction, but rather, it is an instrument (like a telescope) for acquiring intuitions involving induction. It helps us see into Plato's heaven and perceive the right structure. That's putting is rather bluntly and, of course, metaphorically. Perhaps, calling it a form of non-sensory cognition sounds appropriately more tentative.

In Chihara's account, which I quoted above, he uses an interesting expression: 'an intuitive version of mathematical induction'. What could an 'intuitive' version be? The account Chihara gives implicitly assumes that we *already* understand mathematical induction when we gaze upon the diagram, otherwise we would not be able to see induction in it. This suggests that people who have not yet learned induction should be less receptive to the diagram as a proof of the theorem. In my experience with humanities undergraduates, this has not been the case. People with no particular training in mathematical induction grasp the picture proof with no difficulty.

Perhaps a careful experiment could shed light on this. In the meantime, it's interesting to consider how we teach induction in the first place. Like many others, when I teach I use a simple thought experiment involving dominos to convey the principle of mathematical induction. Imagine that I have put a series of dominos on edge, close enough to one another so that if any one should tip over, it will knock over the next one in the series. Now imagine that the first domino in the series is knocked over. What do you think happens to the other dominos? The answer is immediately obvious to everyone: All the dominos tip over.

The thought experiment might involve a pair of important idealizations: that the series of dominos goes on for ever and that it takes no time for one domino to knock over the next. I sometimes mention these idealizations when presenting the example, but it is seldom necessary. Almost everyone gets the idea immediately and goes on to apply the principle of mathematical induction with ease (at least in simple cases). And the distinction between the basis step (domino 1 is tipped over) and the inductive step (if domino n is tipped over, then domino $n + 1$ is tipped over) is completely clear, even to those who loath mathematics and have avoided it since the early stages of their educations when it ceased to be mandatory.

Perhaps the dominos example involves 'intuitive' mathematical induction. In any case, since it teaches induction, it cannot be that someone who is actually learning about induction for the first time from the dominos example recognizes induction in it, or finds induction to be 'suggested' by it. Something else must be going on. I'm inclined to say that the answer involves directing the mind to the abstract realm, but this Platonistic explanation will, of course, be as contentious as my account of the number theory diagram. Chihara and others who claim to see induction in the number theory diagram owe us an explanation of how it is that the dominos example can teach mathematical induction without presupposing it.

In sum, I'm disinclined to think we see induction in the number theory diagram. But it is not too important, one way or the other. There are two points that are important and I wanted to stress them both. One is that the diagram is indeed a genuine proof. Chihara, I'm glad to say, completely agrees. The second point – the one on which we disagree – is that the right account of how the picture works involves some sort of intellectual grasp of abstract entities. And that point stands up, whether the diagram suggests mathematical induction or not.

Three Analogies

My main claim – that some pictures are not really representations, but are rather windows to Plato's heaven – will seem highly implausible to most readers. So, beside the discussion so far, I'd like to add three analogies to make things seems a bit more plausible and palatable.

(1) *From Aesthetics*   Some with an interest in art and psychology distinguish between a 'pictorial' and a 'symbolic' aspect of a representation (e.g. Arnheim 1969). Consider a painting such as David's 'Napoleon'. (It depicts Napoleon; he's in a billowing cape, on a spirited white horse, he's pointing ahead.) As a 'picture' (in Arnheim's terminology), it represents Napoleon; as a 'symbol' it represents leadership, courage, adventure. The painting simultaneously manages to be about something concrete *and* something abstract. It is a wonder that artists can do this, but there is no question that they often do – sometimes brilliantly. (By contrast, my snapshots and doodles are lucky to be pictures.)

What I would like to suggest is that something like this happens with our number theory diagram. It is a *picture* of the special case, $n = 7$, but a *symbol* for every n. Just as we can see courage and adventure depicted in David's painting, so we can see every natural number in the diagram. It's a metaphorical 'seeing', to be sure, but it's a similar sort of perception in each case. If artists can do it, so can mathematicians.

(2) *From Modern Differential Geometry*   In modern presentations of differential geometry and general relativity, geometric objects such as events, vectors pointing from one event to another, tangent vectors, the metric tensor and so on can be characterized independently of any particular coordinate system. Indeed, it is usually easier and more elegant to do so. But these entities can also be given an explicit coordinate representation, and in practice this is often required. The distance between two events in space-time, for example, is an

invariant, an objective feature of the geometry; yet its expression in terms of the location of these events can vary greatly, depending on the coordinate system used.

If we consider a surface, the *intrinsic* features are those which characterize the surface independently of any particular coordinatization. By contrast, the *extrinsic* features depend on particular coordinate systems, and change with a change of coordinates. The connection between them is this: an intrinsic feature corresponds to the existence of a coordinate system with specific appropriate extrinsic features.[7]

The distinction suggests something objective about the intrinsic aspects of a surface, and something less so in the extrinsic, due to the arbitrariness of the various forms of representation. Some concerned with space-time (e.g. Friedman 1983) argue that this distinction corresponds to a distinction between the factual and the conventional; the intrinsic features of space-time (curvature, metric tensor, etc.) are objectively real while extrinsic features are mere artifacts of the form of representation.

The analogy with picture-proofs that I want to suggest is this: Any representation of a surface, say, displaying its curvature, will always be in some particular coordinate system. Analogously, a picture of a numerical relation will always be with some particular number, n. But an intrinsic feature, such as Gaussian curvature, is independent of any particular coordinate system. Similarly, the evidential relation in the number theory diagram is independent of any particular n-element picture. To calculate the Gaussian curvature, however, some particular coordinate system is required. And when we have calculated it in one, we know it in all. Similarly, to grasp the pictorial evidential relation some particular n-element picture is needed. But again, understanding it in one is to understand it for all.

Because of this analogy I'm tempted to call number theory diagrams 'extrinsic pictures', since they are particular representations like particular coordinate systems. Is there such a thing as an intrinsic number theory diagram? Of course. It's the one seen by the mind's eye – and it has no particular number of elements in it. That's the one we ultimately grasp as evidence for the theorem.

(3) *From 'natural kind' reasoning*   One sort of inductive inference is 'enumerative'. We notice that all of a very large number of observed ravens are black; so we infer that *all* ravens are black. This is a commonplace but, often in science, powerful inferences are made from a single case. In high-energy physics, for instance, a single event (sometimes called a 'golden event') captured in a bubble chamber photo will sometimes be sufficient evidence for a powerful general conclusion. One positron is sufficient to generalize about the mass of all positrons. One sample of water is sufficient to establish that all water is H_2O. The form of inference seems based on a principle that runs like this: *If X is a natural kind and has essential property P, then all instances of the*

kind have property P. The assumption at work would then be that positrons or water are natural kinds and that their mass or chemical composition are essential properties. In principle, only one instance is needed to allow us to draw the general conclusion about all positrons or all water. In practice, a few more than one instance are likely to be necessary, simply to build confidence that no mistakes were made in measuring the mass or analysing the composition of the sample. Even so, the power of natural kind inference is remarkable.

Something analogous to natural kind inference is going on in the number theory picture-proof. We can take the diagram $n = 7$ to be an instance of a natural kind; and we further take the formula $n^2/2 + n/2 = 1 + 2 + 3 + \ldots + n$ to be analogous to an essential property of numbers. Since it holds for the particular number $n = 7$, it holds for every n.

The fly in the ointment is this: Water is *essentially* H_2O, but only *accidentally* thirst-quenching. Mathematical objects would seem to have only essential properties, and it would be a terrible inference to pass from '7 is prime', to 'all numbers are prime'. So I'm reluctant to see the inference from pictures to all generality as a clear case of natural kind inference; but it is interestingly similar.[8]

Are Pictures Explanatory?

Mathematicians look for two things in a proof – evidence and insight. Traditionally, a proof must firmly establish the theorem. That, for just about everyone except Lakatos, is a *sine qua non* for any proof. But a good proof also helps us to understand what's going on. Insight, understanding, explanation are somewhat nebulous, but highly desirable. Proofs needn't have them, but are cherished when they do.

Are picture-proofs rich in insight? Many commentators suggest as much; they even play evidence and insight off against one another, suggesting that what we lose in rigour we make up in understanding. (Polya is perhaps the best example of this.) However, this seems slightly misguided. And I would be seriously misunderstood, if it were thought that I am suggesting it is worth giving up some rigour in exchange for insight. This is doubly wrong. I don't see any abandoning of rigour by allowing the legitimacy of picture-proofs. And second, greater insight isn't always to be found in pictures.

In the two number theory cases above, a proof by induction is probably more insightful and explanatory than the picture-proofs. I suspect that induction – the passage from n to $n + 1$ – more than any other feature, best characterizes the natural numbers. That's why a standard proof by induction is in many ways better.

To be sure, some insight is garnered from the diagrams which prove the two infinite series examples. From looking at them, we understand why the series have the sums that they do. Pictures often yield insight, but that is not essential. The examples I have given are mainly a form of evidence – a different form, to be sure, than verbal/symbolic proofs; but they have the same ability to provide justification, sometimes with and sometimes without the bonus of insight and understanding.

So Why Worry?

Philosophers and mathematicians have long worried about diagrams in mathematical reasoning – and rightly so. They can indeed be highly misleading. Anyone who has studied mathematics in the usual way has seen lots of examples that fly in the face of reasonable expectations. I've painted a rosy picture so far, but I'm well aware of pitfalls. Some of these I'll take up in the final chapter.

I realize that talk of 'the mind's eye' and 'seeing mathematical entities' is highly metaphorical. This is to be regretted – but not repented. Picture-proofs are obviously too effective to be dismissed and they are potentially too powerful to be ignored. Making sympathetic sense of them is what is required of us.

Further Reading

There is remarkably little written on this topic. Barwise, J. and J. Etchemendy, 'Visual Information and Valid Reasoning' and And Hammer's *Logic and Visual* Information are good places to start. Also important are various papers of Marcus Giaquinto (see Bibliography). Giaquinto's forthcoming book, *Visual Thinking in Mathematics: An Epistemological Study* will be a major contribution. Mancosu, Jorgensen and Pedersen (eds), *Visualization, Explanation and Reasoning Styles in Mathematics* contains a number of papers addressing visualization. The two books by Nelson, *Proofs Without Words* (vols. I and II), are an excellent source of examples.

Mathematical books now have more and better pictures than ever before. This is largely, though not entirely, due to the marvellous computer graphics which have become available in recent years. This is reflected in three books that I've enjoyed reading: Binmore and Davies, *Calculus*, Needham, *Visual Complex Analysis*, and Francis, *A Topological Picture Book*.

Appendix

Here are some more examples. I've been collecting them for a while, but recently the job was made very easy by Roger Nelsen, whose two books *Proofs Without Words* (vol. I 1993, vol. II 2000) contain a great many gems. Some of the ones above as well as those that follow can be found in his books, which I warmly recommend. In each case, I'll just state the result and present the picture-proof, leaving it to the reader to figure out how the picture works. Remember, pictures may make a result 'obvious', but *obvious* and *immediate* need not be the same thing. Often you will have to work at it for a while.

Given two numbers, *a* and *b*, the *arithmetic mean* is $(a + b)/2$ and the *geometric mean* is $\sqrt{(ab)}$. They are related by an important inequality: $(a + b)/2 \geq \sqrt{(ab)}$. Figure 3.10 gives a proof. There is a bit of algebra that goes with it. Can you also tell from the diagram when equality holds?

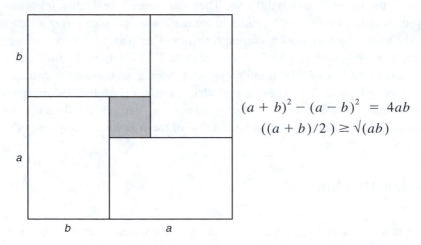

$$(a + b)^2 - (a - b)^2 = 4ab$$
$$((a + b)/2) \geq \sqrt{(ab)}$$

Figure 3.10

Here's a second picture-proof of the same inequality (Figure 3.11).

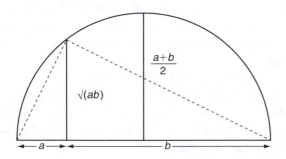

Figure 3.11

Early in one's algebra training, usually before learning the formula for solving quadratic equations, the method of 'completing the square' is taught. Here's the algebra: $x^2 + ax = (x + a/2)^2 - (a/2)^2$. Now here's the proof (Figure 3.12):

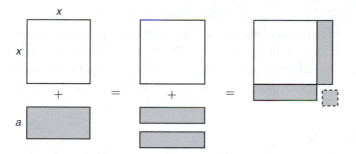

Figure 3.12

As is well known, an arbitrary angle cannot be trisected using only a straight edge and a compass. But it can be trisected by other means. Here's a picture (Figure 3.13) of a mechanical device that will do the trick, as can be seen in a flash.

Figure 3.13

The impossibility of trisecting an angle by straight edge and compass is really an impossibility in a *finite* number of steps. What if you could perform infinitely many operations? Suppose you took one minute for the first operation, then half a minute for the second, then a quarter-minute for the third, and so on. You could carry out infinitely many operations in just two minutes. Could you trisect the angle then? You need to know the following fact about infinite series: $\frac{1}{3} = \frac{1}{2} - \frac{1}{4} + \frac{1}{8} - \frac{1}{16} \ldots$ Now the picture can answer the question.

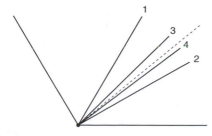

Figure 3.14

Fibonacci numbers are defined as follows: the first is 1, the second is also 1, then each successive Fibonacci number is equal to the two predecessors; so, the third is $1 + 1 = 2$, the fourth is $1 + 3 = 4$, the fifth is $3 + 4 = 7$, and so on. More formally, $F_1 = 1$, $F_2 = 1$, $F_{n+2} = F_{n+1} + F_n$. Here is an interesting theorem about the squares of Fibonacci numbers. $F_1^2 + F_2^2 + \cdots + F_n^2 = F_n F_{n+1}$. Here is the proof.

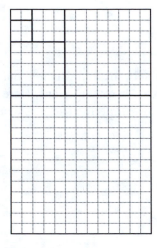

Figure 3.15

CHAPTER 4
What is Applied Mathematics?

We count apples and divide a cake so that each guest gets an equal piece; we weigh galaxies and use Hilbert spaces to make correct predictions about spectral lines. It would seem we have no difficulty in applying mathematics to the world; yet the precise role of mathematics in its various applications is surprisingly elusive. How does it work? Eugene Wigner has gone so far as to say that 'the enormous usefulness of mathematics in the natural sciences is something bordering on the mysterious and that there is no rational explanation for it' (1960: 223). Could he be right?

Some of the issues which arise here are not much discussed under the heading 'Philosophy of Mathematics', much less under the heading 'Philosophy of *Applied* Mathematics', yet they are pivotal to several philosophical debates. Three rather general questions are central:

(1) Just how does mathematics 'hook onto' the world? This is the main concern of a rather technical branch of philosophy of science known as measurement theory.

(2) Are some of the objects referred to in various theories merely mathematical objects or do they have some other status? This problem often comes up in the philosophy of the special sciences. For example, do space-time and the quantum state exist in their own right, separate from their mathematical representations, or are they nothing but mathematical entities?

(3) Is mathematics essential for science? Since Quine and Putnam first said Yes, and were followed by Field who said No, this has become a focal point in the debate between realists and nominalists in the philosophy of mathematics. The debate turns on how mathematics is applied.

Representations

Let's begin by asking how mathematics is applied. The common view in measurement theory begins by assuming two distinct realms: one is a mathematical realm which is rich enough to represent the other, a distinct non-mathematical realm. We pick out some part or aspect of the world and find a similar mathematical structure to represent it. For example, *weight* is represented on a numerical scale. The main physical relations among objects that have weight (determined, say, by a balance beam) are that some have more weight than others and that when combined their joint weight is greater than either of their individual weights. Weight can then be represented by any mathematical structure (such as the positive real numbers) in which there is a *greater than* relation matching the physical greater than relation and an *addition* relation matching the physical addition or combination relation.

More generally, a mathematical representation of a non-mathematical realm occurs when there is a homomorphism between a relational system **P** and a mathematical system **M**. **P** will consist of a domain D and relations R_1, R_2, ... defined on that domain; **M** similarly consists of a domain D* and relations R_1^*, R_2^*, ... on its domain. A homomorphism is a mapping from D to D* that preserves the structure in the appropriate way.

To make this a bit more precise, consider a simplified example. Let D be a set of bodies with weight, let D* = $\mathbb{R}$, the set of real numbers; furthermore, let $\leqslant$ and $\oplus$ be the relations of *physically weighs the same or less than* and *physical addition*. The relations $\leq$ and $+$ are the usual relations on real numbers of *equal or less than* and *addition*. The two systems, then, are **P** = $\langle D, \leqslant, \oplus \rangle$ and **M** = $\langle \mathbb{R}, \leq, + \rangle$. Numbers are then associated with the bodies ($a,b,$ ... in D) by the homomorphism ϕ: D $\rightarrow \mathbb{R}$ which satisfies the two conditions:

(1) $a \leqslant b \rightarrow \phi(a) \leq \phi(b)$
(2) $\phi(a \oplus b) = \phi(a) + \phi(b)$.

In plain English, (1) says that if a weighs the same or less than b, then the real number associated with a is equal or less than the number associated with b, and (2) says that the number associated with the weight of the combined object $a \oplus b$ is equal to the sum of the numbers associated with the objects separately. In other words, the relations that hold among physical bodies get encoded into the mathematical realm and are there represented by relations among real numbers. One of the objects can be singled out arbitrarily to serve as the unit weight, u, so that $\phi(u) = 1$.

I must add a caveat to the assumption of two distinct realms, mathematical and non-mathematical. Since they are linked by the embedding homomorphism, ϕ, which is a function defined on D, there must be *sets* of non-mathematical objects as well as pure sets. This means we start with the usual set theory including *urelementen* or individuals. Among *urelementen* are physical objects, of course, but also abstract and fictional objects (faith, hope and charity are three virtues; Santa's sleigh is pulled by eight flying reindeer). Having sets, sets of sets, etc. of *urelementen* is just a start. The difficult part in setting up or discovering an association between the physical system **P** and some mathematical system **M** usually consists in finding the right set of physical relations. Much of the focus of current measurement theory is in psychology and the social sciences where attempts to quantify such concepts as utility, desirability, IQ, degree of belief, intensity of pain, etc. are exceptionally difficult – and in some cases, foolish.

It must be said – and I'm happy to say it – that this characterization of mathematics heavily favours Platonism, since we are implicitly endorsing the existence of a distinct mathematical realm with which we represent the natural world. Of course, nominalists will reply that we represent with numerals, not numbers, so the point does not conclusively favour Platonism. Still, the naturalness of Platonism in applied mathematics, just as in pure mathematics, is manifestly obvious.

Measurement theory often classifies different types of scale; *ordinal* measurements are the simplest. The Mohs scale of hardness, for example, uses the numbers 1 to 10 in ranking the physical relation of 'scratches'; talc is 1 and diamond is 10. Addition plays no role; the only property of the numbers used is their order, which, by the way, is a strict linear order; each thing scratches or is scratched by each other thing, and nothing scratches itself – at least not in public. By contrast, addition is crucial in *extensive* measurements, such as measurements of weight. (In this case the physical combination of two bodies is represented by the addition of two real numbers. But the embedding homomorphism isn't always so simple as it is in the everyday case of weight; the relativistic addition of two velocities, for example, is constrained by an upper limit on their joint velocity.) An *interval* measurement uses the *greater than* relation between real numbers, but does not employ addition. (Temperature and (perhaps) subjective probability are examples. Two bodies at 50° each do not combine to make one at 100°.)

In passing, it should be noted that the mathematical representation of the world need not be with numbers. From the Greeks to Galileo and after, geometrical objects did the representing. The increasing speed of a falling body, for example, was represented by Galileo by a sequence of increasing areas of geometrical figures. Newton's *Principia* was written in this geometrical style, but thereafter the tremendous power of the calculus has made analysis dominant and pushed geometry into the background. The geometric spirit, however, is far from dead. (See, for example, the 'visual' book by Abraham and Shaw (1983).) Graphs are geometrical, of course, but they tend to depict numerical results; that is, they are representations of representations of the world.

Colour, beauty and other such things are not readily mathematizable. But the alleged subjectivity of these properties has nothing to do with it; felt warmth and pain intensity are subjective, but have the appropriate structure to be mathematized. The reason for the non-mathematizability of colour may have more to do with its internal features; it does not have the same structure as mass, length, temperature or other so-called extensive magnitudes which would make it easy to associate with the real numbers.

So far I have spoken loosely of numbers hooking onto objects. Perhaps, instead of *objects*, numbers are associated with *properties of objects*. From a practical point of view, there isn't much difference, but philosophically the divergence is considerable. The former view is strongly empiricist and dominant today (Nagel 1932, Krantz *et al.* 1971); the later is somewhat Platonistic and has had notable support, too (Russell 1903/1937, Campbell 1920, Mundy 1987, Swoyer 1987). Here I don't mean 'empiricist' or 'Platonist' about numbers, but about the physical world itself. The natural languages for these accounts are first- and second-order logic, respectively. To say that the weight of a and b combined is such and such is to say, according to the first-order theory of measurement, that there is an object c which equals the weight of a and b combined (understood in a somewhat operationalist way with c balancing a and b on a scale). This is physically unrealistic, and at best an idealization; it makes 'the mass of the universe', for example, a very problematic notion. However, it is not a problem for the second-order theory, since it is not objects, but properties that are assigned numbers. The property *weight* is postulated to be continuous and unbounded; there needn't be exemplars of any particular weight in order to talk meaningfully about it.

These two accounts of measurement tie into rival accounts of laws of nature. The relations that hold in the (non-mathematical) relational structure are sometimes, presumably, laws of nature. The empiricist-motivated regularity theory fits harmoniously with the first-order theory. The more realist account of some philosophers which takes a law of nature to be a relation between universals (i.e. properties) fits very naturally with the second-order version.[1] So the question, Does mathematics hook onto objects or onto properties of objects?, may have a bearing on the metaphysical issue of the nature of scientific laws. I won't pursue this issue further. I raise it only to whet appetites. The main point I have tried to establish concerns the *representational* character of applied mathematics: *Mathematics hooks onto the world by providing representations in the form of structurally similar models.*

Artifacts

The second concern with the role of mathematics in the sciences involves the possible presence of mathematical artifacts. Measurement theory is somewhat far-fetched in assuming we can *first* discern relations among non-mathematical objects, then *later* pick out mathematical structures to represent them. In reality, of course, mathematics plays an enormous role right from the start in theory construction.[2] Because of this, it is sometimes difficult to distinguish the mathematics proper from its physical counterparts. For example, the average family has 2.4 children. Of course, there is no family with that many children; the 'average family' is a mathematical artifact. No one is likely to be fooled by this example, but many of the things physicists regularly talk about have a contentious status: Are they physically real, merely mathematical, or what?

I should mention that the expressions 'mere' and 'nothing but' are not meant to be as deflating as they sound. Being a *mere* mathematical entity is not some second-rate status. I would take great pride in being an integer – if that made any sense.

When Maxwell introduced classical electrodynamics, his electrodynamic field was thought by many to be just a mathematical entity. In terms of measurement theory, this is to say that the domain of the physical theory consisted of charged particles, but no fields. This relational structure would then be embedded within a mathematical structure of a vector field. So the only 'field' is the mathematical one. The following argument tipped things the other way: Consider two separated charged particles. If one is wiggled, the other jiggles at a *later* time. During the distinct motions of the two particles all energy can be located in the particles themselves, but not at intermediate times. Energy is conserved and must be located somewhere. Thus, it must be in the field; so the field is physically real. (Note that this argument would not apply to the gravitational field of Newton; action is instantaneous in that theory, so energy can always be located in particles.) The consequence of this is that the electromagnetic field, though it is represented by a mathematical vector field which is isomorphic to it, is a distinct, physically real entity, not a mere mathematical artifact.

Similar problems about how to interpret the mathematical apparatus arise in quantum mechanics and in space and time. Quantum mechanics makes heavy use of a notion of *state*, represented by a vector, Ψ, in a Hilbert space. The mathematics of Ψ is reasonably well understood; the same cannot be said about the state. One view says that there is nothing to the state other than the mathematical vector, Ψ, itself. (Physics texts use the same symbol for both, making this seem natural.) At the other extreme Ψ might be a real field (e.g. Bohm's quantum potential). So much (but certainly not all) of the problem of interpreting quantum mechanics amounts to determining how mathematics hooks onto a quantum system: Is the mathematical vector, Ψ, associated with the electron or with the *state* of the electron?

The modern debate between absolutists and relationalists in space-time concerns the status of the space-time manifold (Friedman 1983). Undisputed is the reality of events. Absolutists hold that actual events are the occupied points of a larger space-time manifold, which is taken to be physically real. (Some prefer to think of space-time points as abstract entities. Whether physical or abstract, however, the main claim is that they are real and distinct from their mathematical representation.) The space-time manifold is then associated with the mathematical structure, $\mathbb{R}^4$. By contrast, relationalists hold that the set of events is directly associated with $\mathbb{R}^4$ (bypassing the manifold). So, once again, a major philosophical issue turns on the question of how mathematics is applied to the world.

There is no general principle or even rule of thumb which will guide us to the right answer when asked: Is this real or is it a mathematical artifact? Each case will have to be made on its own. It is up to the physicists and philosophers of physics to decide, as best they can, which view is correct. The philosophy of mathematics can clarify things in general terms by pointing out the general form of applied mathematics, but it can't hope to pronounce in any particular case on what is and what is not physically real rather than being an artifact of the mathematics.

Bogus Applications

There is a different, but loosely related issue that should be briefly mentioned – the abuse of mathematics, especially in the social sciences. *The Bell Curve: Intelligence and Class Structure in American Life* by Herrnstein and Murray (1996) claims deep connections between intelligence and race and class. To the untutored it may appear a careful, judicious, and – above all – a mathematically rigorous work which makes its case. In reality, it's a vile work fraught with statistical hocus-pocus.[3]

When thinking about cases like this, it's tempting to think: if only people were mathematically better educated – as a condition of citizenship – we'd be protected from this nonsense. But this may be a Utopian hope; perhaps we'd just become susceptible to more subtle fallacies. Often the issues can be very complex and sorted out only by true experts. A recent debate between Herbert Simon and Neil Koblitz is interesting and instructive.[4]

Simon is a very distinguished polymath, famous for work in psychology and computer science, philosophy of science, a leader in artificial intelligence, and a Nobel Prize winner in Economics (1978). Koblitz is a younger, though very prominent, number theorist who has very strong interests in development and mathematical education in the Third World. In short, two heavyweights.

In the 1970s a prominent American political scientist, Samuel Huntington, presented some equations that were intended to characterize various political features such as modernization.

$$\frac{social\ mobilization}{economic\ development} = social\ frustration$$

$$\frac{social\ frustration}{mobility\ opportunities} = political\ participation$$

$$\frac{political\ participation}{political\ institutionalization} = political\ instability$$

When applied to various societies, Huntington concluded, for example, that the Philippines was stable while France wasn't and – most shockingly – that South Africa (then living under apartheid) was a 'satisfied society'. The whole thing smacked of pseudo-science to Koblitz and another mathematician, Serge Lang. The latter raised serious objections to Huntington's nomination to the National Academy of Sciences. And upon Huntington's defeat, the right-wing popular press was outraged. George Will, for example, called on the United States government to withhold NAS funds until the Huntington decision was reversed.

Simon came to the defence of Huntington with 'Some Trivial But Useful Mathematics'. And Koblitz replied at length in *The Mathematical Intelligencer* in an opinion piece called 'A Tale of Three Equations; Or The Emperors Have No Clothes' (1988). It was followed by a reply by Simon, and subsequent replies by both and other mathematicians in the next two issues. The debate turned on several questions, including Simon's characterization of key concepts such as 'monotonic', 'continuity' and so on. Koblitz summed up the work of Huntington and Simon, saying: 'Mathematical verbiage is being used like a witch doctor's incantation, to instill a sense of awe and reverence in the gullible or poorly educated' (Koblitz 1988: 10). And further, 'as the South African example shows, the type of pseudo-quantitative methodology promulgated by Huntington and Simon can be as pernicious as it is scientifically vacuous' (*ibid.*: 16).

Though Simon would not admit defeat, it was clear to most mathematically sophisticated readers that game, set and match went to Koblitz. The example is instructive in two ways. It illustrates the bogus use of mathematics in pseudo-science. But it also shows how difficult things can get. Clearly we would all be better off if everyone were mathematically better educated. Then we could defend ourselves from parapsychologists, IQ-hucksters, advertisers and unscrupulous politicians. But the amount of mathematics needed to protect oneself from Huntington's bogus equations and their elaboration and defence by Simon is extensive. There is probably no hope for us here but to rely on experts. Professional mathematicians have a civic duty – however distasteful they may find applied mathematics – to become involved.

By the way, more recently Huntington published the best seller, *The Clash of Civilizations*. This time the mistakes are not mathematical.

Does Science Need Mathematics?

Let's now take up the third question: Is mathematics necessary for science? The answer may be Yes, but it is not obviously so. The statement 'There are two apples in the basket' seems to make essential use of numbers; yet we can capture its content without any appeal to mathematics at all by recasting it as: $\exists x \exists y \forall z(Ax \ \& \ Ay \ \& \ x \neq y \ \& \ (Az \rightarrow z = x \lor z = y))$, where 'A' means 'is an apple in the basket'. Hartry Field (1980) maintains that all of science can be done – in principle – in the spirit of this simple example *without the use of numbers*. Of course, there's no denying that mathematics is heuristically powerful and perhaps even psychologically essential for doing the physics that has been done to date, but, according to Field, it is not necessary in any deep ontological sense.

Field is mainly interested in combating a view of Quine and Putnam. They claim that since mathematics is essential for science, it must be true; and since it's true, there must exist such objects as sets, functions, numbers, etc.

> [Q]uantification over mathematical entities is indispensable for science
> . . . therefore we should accept such quantification; but this commits us
> to accepting the existence of the mathematical entities in question. This
> type of argument stems, of course, from Quine, who has for years
> stressed both the indispensability of quantification over mathematical
> entities and the intellectual dishonesty of denying the existence of what
> one daily presupposes.
>
> (Putnam 1971: 57)

The term 'quantify over' means to assert the existence of, as in 'There are apples in the basket' or 'There is an even prime number'. Quine's criterion, which Putnam endorses, says that if we hold such statements true and there is no way to paraphrase away such expressions, then we are committed to the existence of (in this case) apples and prime numbers. The upshot for Putnam is that there is no paraphrasing away of number talk – mathematics is essential for science. Thus, honesty demands that we acknowledge that numbers and other mathematical entities really do exist.

Against this Platonism Field upholds a brand of nominalism, claiming that mathematics is not essential, but only provides an extremely useful short cut.[5] Field claims, in particular, that the role played by mathematics is quite different from that played by other theoretical entities such as electrons. Field is surely right about this last point: mathematics works, as we argued above, by providing models in which the world (or some part of it) is *represented*. (But this, of course, does not mean Field is right in his nominalism. There are many other – much better – reasons for mathematical realism than the one he attacks.)

In this representing capacity, says Field, mathematics is *conservative*. His principle result is this: If A is a consequence of $T + S$ (where T is a nominalistically acceptable theory and S is a mathematical theory), then A is a consequence of T alone. (Schematically, see Figure 4.1.) The conservatism claim is then used by Field to justify his view that mathematics is not essential for science, since the consequences of the theory exist independently of mathematics.

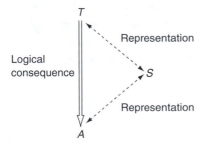

Figure 4.1

Field's commentators have been largely sceptical (e.g. Hale (1987), Irvine (1990), Malament (1982), Shapiro (1983a, 1983b), Tiles (1984), Urquhart (1990) and, most recently, Colyvan (2001)). Among the objections have been these: The notion of *logical consequence* that is needed is that of second-order logic. But second-order logic is not recursively axiomatizable, which means that the notion of consequence must be semantical. Syntactical consequence (i.e. a derivation) is perhaps nominalistically acceptable (arguably being just a string of symbols), but surely not semantical consequence, since this involves the idea of being *true in all models*, a set-theoretic idea if ever there was one. (Quine (1970) famously holds that second-order logic is really just set theory in disguise.)

It would seem, moreover, that we need mathematics to make sense of some crucial notions like *determinism*. In doing physics we talk not only about how things *are*, but about what is or isn't *possible*. For example, determinism can be defined as follows: A theory is *deterministic* if all of its models with the same initial conditions have the same final conditions; it is *indeterministic* when two of its models with the same initial conditions have different final conditions. Obviously, for this we need the notion of a model, and typically that is the kind of abstract entity provided by set theory that seems to give the nominalist indigestion.

Perhaps concern with determinism is not really part of science proper but is instead a purely philosophical issue. In that case, mathematics would seem to be essential for metaphysics – an amusing thought.

Related to this sort of consideration, though less precise, is the role of mathematics in methodology. Empiricists in the past have often maintained that the meaning of a theoretical term (electron, gene) must be given via observation terms. Most philosophers today have abandoned this view, leaving it something of a mystery how we do manage to understand highly theoretical notions. Mathematics may provide the answer since it would seem to provide a framework for thinking about the world. Highly theoretical concepts (symmetry, resonance, isospin, etc. in particle physics) which have no hope of being tied directly to empirical concepts, can often be easily explicated and understood via mathematical models. This would seem to make mathematics not just heuristically useful in drawing consequences from our scientific theories (as Field readily grants), but *methodologically essential* in the very creation and comprehension of those theories – it's crucial for concept formation.

Finally, though this is a tangential point, it may be a mistake to think that theories are nominalistically acceptable, independently of the mathematics. The example that Field develops in detail is Newtonian gravitation theory. This involves massive bodies and space-time points. These are probably acceptable for a nominalist, though some critics have objected that even space-time points might be abstract. (The issue is complicated, since space-time points are neither clearly inside nor outside space-time.) However, some theories employ abstract entities right from the start. As mentioned above, the quantum state, Ψ, for instance, is arguably not just a mathematical entity, but a real (though abstract) object with something like causal powers of its own.[6] And if the best account of laws of nature involves the postulation of relations among properties (universals), then nominalism is hopeless, anyway.

Representation vs. Description

At the outset I made the assumption that there are two, quite distinct realms: the mathematical and the non-mathematical, and that in applications the former *represents* the latter. This isn't the only way to view the situation. Perhaps mathematics *describes* the world. The Pythagoreans, for example, thought that the world *is* mathematical. And John Stuart Mill held that numbers are a kind of very general property that objects possess. A four-legged, blue, wooden chair has the property *four* just as it has the properties *blue* and *wooden*. Philip Kitcher (1983) has proposed an updated version of Mill. Elementary arithmetic, for example, stems from our ordinary experience; such statements as $2 + 3 = 5$ are not truths about a separate mathematical realm, but are rather general truths about the physical world. More sophisticated mathematics is created by an 'ideal agent' who can carry out infinitely many operations. The application of mathematics to the world is, consequently, no more mysterious than is the applicability of such high-level generalizations as 'Red and yellow mixed together make orange'.

Like Mill, Kitcher goes far in explaining how mathematics is applied to the world. To be fair, though, many of the sophisticated uses of mathematics don't seem to fit this view. The *properties* of quantum systems, for example, are associated with the eigenvalues of linear operators defined on a Hilbert space. It is wholly implausible to see this as an extension from everyday experience exemplified by counting bananas. Perhaps the greatest weakness for Kitcher, as it was for Mill, is in doing justice to pure mathematics. Kitcher posits an 'ideal agent', for example, who makes indefinitely complicated calculations. This idea – interesting though it is – has been repeatedly criticized for being an assumption just as strong as any made by Platonism, and a good deal more obscure.

Looking back on the debate, Field versus Quine and Putnam, we can see it as an implicit debate about whether mathematics *represents* (Field) or *describes* (Putnam and Quine). I think of Field as being on the representational side since he has explicitly used the results of measurement theory as done in the representational way. My reason for saying Quine and Putnam see mathematics as describing the world stems from their various remarks about the possibility of revising mathematics and logic in the face of experience. On the representationalist (or modelling) view of applied mathematics, this would be absurd, since an empirical upset would simply make us look for a different mathematical model to represent things; it would not lead us to change our mathematical theories themselves.

To see this in a simple case, consider the addition of velocities. Imagine a ball thrown with velocity W inside an airplane which is flying at velocity V with respect to the ground. We associate real numbers with these velocities: $\phi(W) = w$ and $\phi(V) = v$. In classical physics the composition of velocities, $\oplus$, takes a simple form: $\phi(W \oplus V) = \phi(W) + \phi(V) = w + v$. However, in relativistic physics the composition of velocities is more complicated: $\phi(W \oplus V) = (\phi(W) + \phi(V))/(1 + (\phi(W) \times \phi(V))/c^2) = (w + v)/(1 + (wv)/c^2)$.

Obviously, this was not an overthrow of our previous beliefs about mathematical addition. Indeed, the old mathematical '+' plays a role in the new formula – it's still addition. Rather, we have simply picked out a different mathematical structure on which to model the physical composition of velocities.

It must be said that the entire history of mathematics very strongly supports the autonomy of mathematics and hence strongly supports the representationalist account. Mathematical results have been overthrown, but always by other bits of mathematics. Results in one part of physics have sometimes led to a revolution in other parts, even to revolutions elsewhere, such as in chemistry. But *never* has a result in physics led to the overthrow of any result in the mathematical realm. The discovery of non-Euclidean geometry, for example, was a mathematical discovery. Once the existence of such geometries was recognized, it allowed the possibility of new ways to represent or model the physical world. The success of such new representations (i.e. General Relativity) stimulated in turn further work in differential geometry. But the connection between the mathematical theory and the physical theory is *psychological* – not logical; developments in one provoke an interest in the other. On the other hand, the older chemical views were *logically* refuted by quantum mechanics. Nothing like that has ever happened in mathematics. This epistemic autonomy argues rather decisively for ontological autonomy, and hence for the representational rather than descriptive nature of applied mathematics.

Let's turn now to another descriptivist view that's been gaining popularity in recent years.

Structuralism

In the campaign for the affection of realists, Platonism's chief rival is structuralism. Both are realist views in that they take mathematical statements to have a truth value that is in no way dependent on us, but a truth value that is nevertheless discoverable. The principal difference is how each views mathematical objects. Platonists think of mathematical statements as describing mathematical objects, just as the statements of science describes the objects of the natural world. (And, as I have argued above, they think of mathematical statements as representing, but not describing, the objects of the natural world.) Mathematical structures are built up out of these objects just as houses are built out of bricks. Structuralists turn this around: for them, structures are primary; mathematical objects are nothing but places in a structure.

Anti-realists will hardly notice a difference between structuralists and traditional Platonists. Indeed, one structuralist (Resnik) explicitly calls himself a Platonist. Nevertheless, the differences are important and interesting, so the terminology – Platonism versus structuralism – will be maintained in the following discussion. In terms of applied mathematics the difference is simple: structuralists hold that mathematics applies directly to the physical world, something like a description; it is not transcendent and representational as the Platonist would have it.

The main champions are Michael Resnik (1981, 1982, 1988, 1997), and Stewart Shapiro (1983a, 1983b, 1989, 1997).[7] A modal version has been proposed by Hellman (1989). They trace their lineage to a paper by Benacerraf (1965), and pay ultimate homage to Dedekind (1888). Naturally, there are subtle differences among them, but the core idea is nicely expressed by Resnik:

> In mathematics, I claim, we do not have objects with an 'internal' composition arranged in structures, we have only structures. The objects of mathematics, that is, the entities which our mathematical constants and quantifiers denote, are structureless points or positions in structures. As positions in structures, they have no identity or features outside of a structure.
>
> (Resnik 1981: 530)

On this view, there are underlying structures which may be common both to the physical world and to mathematical systems. An infinite string of stars, for example, has the same underlying *structure* as an infinite sequence of moments of time, or as an infinite string of strokes, | | | | | | | . . . . It is easy to see why on this view mathematics is applicable to the non-mathematical realm: mathematics describes the structure or pattern, and the structure is present in the physical system itself.

An object is just a place in a structure; a star, a temporal moment, or a stroke could exemplify the number 27 just by being at the appropriate place. There is no number 27 over and above that specific place in some appropriate structure. Like traditional Platonism, structuralism is a realist view of mathematics. But the difference, to repeat, when it comes to applied mathematics is that structuralism sees structures right in the non-mathematical world itself, making mathematics descriptive, while Platonism sees the mathematical world as transcendent, making mathematics representative.

There seem to be three motivations for the structuralist account of mathematics. One has to do with the ontology of mathematics. There are many ways, for example, of characterizing the natural numbers (Benacerraf 1965). One way is to follow Zermelo, defining the natural numbers:

$$0 = \phi, \quad 1 = \{0\}, \quad 2 = \{1\}, \quad 3 = \{2\}, \quad \ldots$$

Another way is to follow von Neumann:

$$0 = \phi, \quad 1 = \{0\}, \quad 2 = \{0,1\}, \quad 3 = \{0,1,2\}, \quad \ldots$$

Which is the right way? This turns out to be very likely an unanswerable question, so the natural response is to say that both are, and more generally that *anything* with the right structure could be the natural numbers, or rather, *is an instance* of the natural numbers.

The other two motivations for structuralism are related to each other and focus on the epistemology of mathematics: How is mathematics applied to reality?, and How do we come to know mathematical truths? The difficulty, as structuralists see it, stems from taking Platonism's abstract objects to be transcendent, hence inaccessible, hence unknowable. Shapiro remarks:

> [T]o hold that mathematics is about a non-physical universe, to emphasize the independence of this universe from the material world, and to *leave it at that* is to ignore and even to obscure one of the most important aspects of mathematics, its importance in scientifically understanding the non-mathematical world. ... the main advantage of structuralism is that it provides a more holistic view of mathematics and science, and this accounts for the rich interplay between the fields.
> (Shapiro 1983b: 532, 541)

Since this chapter is about how the transcendent mathematical realm hooks onto the non-mathematical realm, it does in fact answer Shapiro's *don't-just-leave-it-at-that* criticism. And as for the 'rich interplay', as I mentioned briefly above, one has to be quite careful about what counts as part of the interplay between mathematics and science. Since they do not interact the way physics and chemistry interact, we should be rather guarded in proposing a more 'holistic' account.

Epistemology is a big worry for structuralists. Since I dealt with the *unknowability-in-principle-of-abstract-objects* objection in Chapter 2, I am not inclined to take this type of motivation for structuralism seriously here. But it would be useful to spend some time on structuralist views on this issue.

We see arrangements of physical objects. By means of this sense perception, say structuralists, we are able to know elementary patterns. Of course, one might wonder whether we grasp the pattern by seeing it in the physical objects or instead independently grasp the pattern and impose it on the physical world. The duck/rabbit outline exists independently of us, but it is the mind that imposes one pattern or the other on the outline.

As for much more complicated patterns that are not perceptually evident (e.g. infinite structures), structuralists say these might be conjectured, then accepted on the basis of having the right consequences. This is an appealing idea; Gödel, as I mentioned in an earlier chapter, claimed that this is the reason we accept many axioms of set theory. It would seem to be analogous to something that happens regularly in physics and elsewhere. Unobservable quarks are first postulated; then the theory is tested and ultimately accepted on the basis of having the right observable consequences. But notice that before proposing the quark theory, the concept of quark is first conceived. If structuralists are right, this cannot be what is going on in the case of mathematics. A particular conjecture could be that X has structure S. But to make such a conjecture we must *already* have conceived the idea of the structure S, itself. We cannot be discovering structures by conjecturing them. We can only discover that something has that particular structure, just as we discover that the world actually has instances of the concept quark.

With this in mind, we can readily see the difference between this proposal and that of Gödel. Gödel is merely conjecturing new propositions, not new concepts (or, if you prefer, he is conjecturing that a concept has instances). The axioms of set theory all use just two primitive concepts, *set* and *member*.

Shapiro has made a somewhat different proposal for coming to grips with the epistemic difficulty. He suggests that structures and language are intimately connected; we grasp a particular structure by understanding the language. To be fair, he readily admits this is only a sketch. But even in sketchy form, I think there is a serious objection to it based on mathematical practice.

We often understand words (especially in vague or ungrammatical sentences) because we have independent access (e.g. sense perception) to that which the words refer. Example: Suppose I'm told 'The Englishman in the corner with a beer is a friend of Mary's.' Actually, he's Irish, he's not in the corner, but more towards the middle of the room, and he's drinking lemonade. Nevertheless, the inaccurate description still allows me to approach this fellow and enquire about our mutual friend Mary. The independent sense perception, however, is obviously crucial. It works in conjunction with the (faulty) description. Real working mathematical language is very informal. Our grasp of a particular structure *cannot* depend entirely on this language. It would seem that we have some independent grasp of the structure – either (a) by 'seeing' the structure directly, or (b) by 'seeing' the (mathematical) objects which give rise to this structure. In either case, structuralism is certainly not better off than traditional Platonism from an epistemic point of view.

Let's turn away from epistemic considerations to some ontic problems. Shapiro remarks 'A *structure* is the abstract form of a system, focussing on the interrelationships among the objects, and ignoring any features of them that do not affect how they relate to other objects in the system' (1989: 146). This is very natural when one considers examples such as the baseball infield. We pay attention to the number of players, to where they typically stand, to what they wear in the way of gloves, and so on. We ignore their eye colour, batting average and other things which seem strictly inessential to playing defence. The distinction between essential and the accidental properties springs readily to mind; but such a distinction is not available in mathematics where *all* properties are essential, if any are. (See the discussion, with caveats, of this issue in the previous chapter.) In the set-theoretic definition of the natural numbers, which properties can we ignore as not essential, even in an intuitive sense of this notion? Shapiro suggests 'ignoring any features of [the objects in question] that do not affect how they relate to other objects in the system'. But in the set-theoretic definition of numbers the membership relation explicitly holds or fails to hold between every pair of numbers. Thus in Zermelo's definition we have $2 \in 3$ (i.e. $\{\{\phi\}\} \in \{\{\{\phi\}\}\}$), while $2 \notin 4$. In von Neumann's definition we have any number is a member of every larger number; so, for example, $2 \in 4$ (i.e. $\{\phi, \{\phi\}\} \in \{\phi, \{\phi\}, \{\phi, \{\phi\}\}, \{\phi, \{\phi\}, \{\phi, \{\phi\}\}\}\}$).

The membership relation is well defined over all these sets; that is, it is something that relates every set to every other in the structure. Thus, it cannot be ignored or abstracted from. And there's no denying that membership is important in set theory – it's as basic as anything can be. The membership relation is different in the Zermelo structure from what it is in the von Neumann structure. It would seem, therefore, that they are different structures, after all.

Yet another problem stems from what appears to be an obvious mathematical truth: *Every mathematical structure has at least one instantiation.* In the physical realm this needn't be true. Imagine the game 'super-duper baseball' which is played on a field much larger than the surface of the earth with 100 billion players on each team. The infield is properly characterized in a very complicated way, so complicated that there isn't enough paper for me on which to write it all down. Here is a structure which, it seems safe to say, will never be instantiated; no one will ever play super-duper baseball. Yet we still make perfectly good sense of such a structure. Why? Because the structure is an abstract entity; it needn't be physically real. But in mathematics, we can't have an uninstantiated theory. If a particular (alleged) mathematical structure has no instantiation, then it is because its description is actually inconsistent.

One last problem. Set theory is at the very heart of mathematics; it may even be all there is to mathematics. The notion of set, however, seems quite contrary to the spirit of structuralism. A set is a collection of objects. The complete identity of any set is tied up with is members and with nothing else. In group theory, by contrast, the notion of group has some sort of priority over particular instances of groups. Groups very nicely fit the structuralist account. But sets don't seem like this at all. The members (objects) have a kind of priority over the sets (structures) that they constitute. If a set was just a structure, then changing its members would not effect it any more than changing the first baseman changes the structure of the infield. But a set's identity is wholly dependent on its members – change the member and you change the set. Structuralism doesn't do justice to this basic fact about set theory (see Parsons 1990).

The real upshot of all this is that the support for a descriptive view of mathematics offered by structuralism cannot be upheld and the implicit challenge to the representational account can't be sustained. Platonism still has the upper hand.

In conclusion, let me return to Wigner. 'The miracle of the appropriateness of the language of mathematics for the formulation of the laws of physics is a wonderful gift', he says, 'which we neither understand nor deserve' (Wigner 1960: 237; see also Steiner 1989). This seems very far from the truth. We may not grasp it fully, but we're well on the way; and what remains unexplained is well within the scope of Platonism's representational account.

Further Reading

To develop a feel for this topic, it would be wise to read widely in the sciences that make significant use of mathematics: physics, of course, but also many others as diverse as economics, linguistics, social statistics, and so on. For those who would like to see more of measurement theory, the chief reference is the three volume work by Krantz *et al.*, *Foundations of Measurement Theory*.

Philosophers have focussed on the topic of indispensability. The debate was started by Quine and Putnam; Field replied. The subsequent literature is vast. Colyvan's *The Indispensability of Mathematics* is an excellent recent work on the topic and contains a full bibliography. Wigner's famous remark about the miracle of applied mathematics has been addressed (with religious overtones) by Steiner, *The Applicability of Mathematics as a Philosophical Problem*.

Though it is seldom explicit, Philosophical work on models in science is often about the nature of applied mathematics.

CHAPTER 5
Hilbert and Gödel

The Nominalistic Instinct

O f course, there are red things, but is there *redness* itself? Some people are wise, but does *wisdom* exist in its own right? Many think the answer to these questions obvious: No, such wierd entities do not exist. Those who dismiss them are the nominalists at heart. Abstract terms, according to nominalists, are not the names of abstract objects. Redness and wisdom are just words and nothing more – hence 'nominalism'. As for mathematics, the instinctive nominalist holds that there are no numbers, only numerals. Platonists think that the numeral '2' is the name of the number two, just as 'Jim' names me. But, for the nominalist, there are no numbers; the real subject matter of mathematics is numerals, symbols, and words, all of them strictly meaningless – not in the sense of gibberish, but in the sense that there is nothing that they mean, or name, or to which they refer.

Quite aside from philosophical sensibilities of a nominalist sort, the history of mathematics over the last century and a half might incline one towards conventionalism and formalism. Great advances in geometry, for example, came from looking into the deductive relations among the postulates. By reinterpreting terms such as 'plane', 'straight line', etc. non-Euclidean geometries (which deny the parallel postulate) were shown to be consistent. The rise of abstract algebra (groups, rings, Boolean algebras, etc.) early in this century further re-enforced this practice. Naturally enough, the striking successes of this approach might lead one to think that mathematics is nothing but the study of uninterpreted systems – just symbol manipulation.

Nominalist scruples are mainly negative – they rule out abstract entities. There is still a diverse spectrum of possible views which are compatible with nominalism; some of the more prominent are lumped together in this chapter. Perhaps the most prominent and influential nominalist today is Hartry Field, whose views were examined briefly in the last chapter. Other accounts of mathematics which are akin to nominalism and formalism include: linguistic conventionalism and Wittgenstein's radical conventionalism. But we won't stop to examine these now; instead we will save most of our energy in this chapter for the most brilliant formalist of all, David Hilbert. But first, a very quick look at pre-Hilbert formalism.

Early Formalism

Chess and other board games consist of symbols or tokens and of rules for moving them around. No one would take the chess pieces as denoting anything. Formalists love the analogy: mathematics is just a game; mathematical objects are like chess pieces and mathematical rules are like the arbitrary rules of a game. With a grand flourish they might add that mathematics is the greatest game ever played, but it is just a game, nevertheless.

Consider a simple system, which I'll call S.

Primitives of S: Individuals: ♣,♥;
Properties: ♦,♠

Axioms of S: (1) $\forall x(\blacklozenge x \rightarrow \spadesuit x)$
(2) $(\exists x \spadesuit x) \rightarrow \blacklozenge \clubsuit$
(3) $\spadesuit \heartsuit$

Rules of inference: MP (*modus ponens*), UI (universal instantiation), EG (existential generalization).

Theorem: ♠♣
Proof: ♠♥ (Axiom 3)
∴ $\exists x \spadesuit x$ (By EG)
∴ ♦♣ (By axiom 2 and MP)
∴ ♦♣→♠♣ (By axiom 1 and UI)
∴ ♠♣ (By MP)

The symbols ♥, ♦, ♠, and ♣ are perfectly meaningless. The axioms are not 'true', but are rather stipulations, implicit rules for manipulating the symbols. Except for the fact that it is vastly more complicated and sophisticated, all of mathematics is just like this, according to formalists; it's a game played with ink, chalk and (since we also talk mathematics) sound waves.

Influential nineteenth-century formalists included the mathematicians Thomae and Heine who developed the view extensively. The logician Schröder contributed to the view as well, sometimes amusingly. Since mathematics is about symbols, Schröder felt obliged explicitly to postulate that when his book was closed, the ink marks didn't rearrange themselves into different symbols.

The whole school of (early or pre-Hilbert) formalism was crushed by Frege (1884). I'll only mention two of Frege's points. First, mathematics couldn't possibly be about individual symbols (tokens) but must instead be about classes of symbols (types). The string: ♠ ♠ ♠ ♠ ♠ consists of five distinct *tokens* all of the same *type*. Tokens are concrete individuals, but types are abstract. The tokens may themselves be meaningless symbols, the very thing that nominalists hanker after, but types will give them heartburn. If we're going to have abstract objects anyway, we might as well hang for an eggplant as an okra.

Second, the meta-theory of games can be meaningful mathematics. (This, perhaps, was Frege's strongest point.) For example, consider: 'The proof of the (above) theorem is five steps long.' 'The king and two knights cannot force mate.' These are meta-theorems about games – *and they are meaningful*. The game of chess and the formal system S are perhaps meaningless games in the very sense that formalists claim. But the meta-claims about those meaningless games are themselves not meaningless. They are mathematical and they have an objective truth-value.

Hilbert's Formalism

David Hilbert (1862–1943) was a brilliant German mathematician, one of the greatest ever, and along with Poincaré, the dominating figure of the late nineteenth and early twentieth centuries. He made spectacular contributions to number theory, analysis, geometry and theoretical physics, as well as to philosophy and foundations of mathematics. Having been born and raised in Königsberg, he became quite familiar with Kant's views through osmosis as well as explicit study.[1] The Kantian element of Hilbert's view is what separates his formalism from earlier, implausible accounts.

The Kantian background will loom large in constructive accounts of mathematics discussed in a later chapter; here I'll only sketch the main features. According to Kant, objects and events are not 'out there' in space and time waiting for us to experience them. Rather, space and time are contributions from within; they are the *forms of our perceptions*. We experience objects as being in space and time, but space and time do not objectively exist, independently from us – we, in some sense, create them. Geometry is associated with our intuitions of space, and arithmetic with our intuitions of time – numbers are successive like a sequence of events in time.

Kant's view is inherently finitistic in the sense that we obviously cannot experience infinitely many events or move about infinitely far in space. Of course, there is no upper bound on what we can do: no matter how far we move, we can always move a step further, and no matter how many events we have experienced, we can always experience one more. But at any point we will have acquired only a finite amount of experience and have taken only a finite number of steps. Thus, for Kant, the only legitimate infinity is the so-called *potential* infinity, not the *actual* infinity. And, because space and time (or geometry and arithmetic) are our creations, we can know their properties a priori. So, a Kantian like Hilbert would say that finitistic mathematical truths, intimately bound up with perception, can be known with complete certainty.

Hilbert adopted this Kantian framework. His innovation was to apply it to the symbols of mathematics themselves, considered as objects of experience. Consider successions of perceptual objects, such as: $|||||$. It is perceptually evident that the series $|||$ and the series $||$ when put into concatenation yield the series $|||||$. We can abbreviate this by writing $3 + 2 = 5$. It is a certain truth – not a linguistic one, nor a truth about an independent Platonic realm. It is a perceptual truth – not a particular one, such as the perception that the grass on my lawn is green – but an a priori truth about the very structure of *any possible* perception.

If we were content with finitistic mathematics, then this could be the end of the matter. Hilbert, relying on Kant, would have explained and justified all of mathematics. But he wanted very much more than this – and rightly so. He wanted to preserve all of classical mathematics with its infinite totalities. This includes transfinite set theory, about which Hilbert famously declared: 'No one shall drive us out of the paradise that Cantor has created for us' (1925: 191).

Consider the natural numbers, 0, 1, 2, 3, ... Often we say that there are infinitely many of them, meaning: no matter how far we count, we can always count one more. But set theory actually says something much stronger than this; it says that the set of natural numbers, $\omega = \{0, 1, 2, 3, \ldots\}$ is an actual infinite set. This is necessary, for example, in order to make sense of the power-set axiom which says that the set of all subsets of ω also exists, and to make sense of the theorem which declares that the power-set is larger than its underlying set. Thus, the cardinality of $\wp\omega = \{0, \omega, \{0\}, \{1\}, \ldots \{0,1\}, \{0,2\}, \ldots\}$ is greater than the cardinality of $\omega = \{0, 1, 2, \ldots\}$. Platonists, of course, have no trouble with actual infinities while constructivists such as Kant and others (as we shall see in a later chapter) reject them outright, allowing only potential infinities.

Historically, paradoxes and conceptual problems of mathematics have usually stemmed from the infinite. This includes, for example, Zeno's paradoxes in Greek times, infinitesimals in the seventeenth century, and the paradoxes of set theory in the late nineteenth and early twentieth centuries. In every case the problem stemmed from trying to reason with infinite quantities.[2]

Hilbert surveyed the physical realm, but, of course, found no actual infinite totalities there. Given the correctness of atomism, there are no infinitely small things, no infinite divisibility. And given General Relativity, the universe is only finitely large, so there aren't, for instance, infinitely many stars. Thus, the infinite can never be part of our perceptual experience. Hilbert's problem, as he saw it, lies in how infinite mathematics can be incorporated into the finite Kantian framework. We want to keep the extraordinary beauty, power, and utility of classical mathematics, but we also want to incorporate it in such a way that we can have full confidence that no more paradoxes will arise. 'The goal of my theory', says Hilbert, 'is to establish once and for all the certitude of mathematical methods' (1925: 184).

Ultimately, it proved unworkable; nevertheless, Hilbert's solution is ingenious. He, in effect, divides all of mathematics into two parts. The finitistic part is true and meaningful (for the Kantian reasons mentioned above). The infinitistic part is strictly meaningless, lacking any truth-value. Thus, '$3 < 5$' is true (and '$5 < 3$' is false) while '$\omega < 2^{\omega}$' is strictly neither true nor false. In spite of their meaninglessness, statements involving the infinite can be added to meaningful, finite, true mathematics as supplements to make things run more smoothly or to derive new finite results. In short they are adopted for their instrumental value.

Hilbert thought of these as 'ideal elements'. In projective geometry, for example, there are a lot of *almost*-theorems: almost every pair of lines intersect at a point; the exceptions are parallel lines. However, with the introduction of a *point at infinity*, even parallel lines can intersect. The inclusion of this ideal element, a point at infinity, eliminates the need to state exceptions to theorems. In general, it simplifies theorems and proofs to such an extent that ideal elements are a standard part of projective geometry.[3] Points at infinity are taken to be mere fictions, justified by their enormous power and utility.

Of course, we can't go around introducing ideal elements wherever we like. 'There is', Hilbert declares, 'one condition, albeit an absolutely necessary one, connected with the method of ideal elements. That condition is a *proof of consistency*, for the extension of a domain by the addition of ideal elements is legitimate only if the extension does not cause contradictions' (1925: 199).

Hilbert thought of complex numbers and infinite sets as further examples of ideal elements, both profoundly useful because of how they streamline mathematics: 'we conceive of mathematics to be a stock of two kinds of formulas: first, those to which the meaningful communications of finitary statements correspond; and secondly, other formulas which signify nothing and which are the *ideal structures of our theory*' (1925: 196). In short, he held that *classical mathematics = finite mathematics + ideal elements*.

There is a popular account of natural science which is similar to Hilbert's outlook. The analogy is worth considering. Instrumentalists (e.g. Duhem 1906) hold that the statements of, say, physics are of two kinds – observable and theoretical. 'There is a white streak in the cloud chamber' is an observation sentence that can be directly confirmed by experience. On the other hand, 'Electrons are deflected by a magnetic field' is theoretical and is said by the typical instrumentalist to be neither true nor false. We adopt the electron theory, on this account, not because it is true, but because it is extremely useful in organizing and predicting observation sentences. In general, theories are not attempts to describe the world truly; they are mere instruments for predicting observations. (By contrast, so-called scientific realists say that theories are attempts to give true descriptions of reality.) Hilbert, then, is like the instrumentalist: finite, meaningful mathematical statements are like the observation statements of natural science, while the infinite parts (the ideal elements) of mathematics work like theoretical entities understood as useful fictions.

Continuing the analogy, we can now see how Hilbert's central problem arises. Suppose we have two theories, a theory of heat, H, and a theory of light, L. A scientific realist takes evidence for H and evidence for L to be evidence that they are each *true*. Given the truth of H and L separately, belief in the truth of the *conjunction* H & L follows on naturally. But the instrumentalist cannot be so sanguine, since evidence is understood merely as evidence that a theory is a good instrument. Thus, evidence that H is a good instrument and evidence that L is a good instrument need not be evidence that H & L is also a good instrument. For even though H and L are individually consistent (a precondition of being a good instrument), their conjunction need not be. What Hilbert needs to do is to show that the various parts of infinite mathematics will fit with one another and with finite mathematics in such a way that no inconsistency can be derived.

But what is involved in deriving things? Indeed, what is involved in mathematical reasoning in general? Hilbert fixes on the symbols themselves.

Does material logical deduction somehow deceive us or leave us in the lurch when we apply it to real things or events? No! Material logical deduction is indispensable. It deceives us only when we form arbitrary abstract definitions, especially those which involve infinitely many objects. In such cases we have illegitimately used material logical deduction; i.e. we have not paid sufficient attention to the preconditions necessary for its valid use. In recognizing that there are such preconditions that must be taken into account, we find ourselves in agreement with the philosophers, notably Kant. Kant taught – and it is an integral part of his doctrine – that mathematics treats a subject matter which is given independently of logic. . . .

As a further precondition for using logical deduction and carrying out logical operations, something must be given in conception, viz., certain extralogical concrete objects which are intuited as directly experienced prior to all thinking. For logical deduction to be certain, we must be able to see every aspect of these objects, and their properties, differences, sequences, and contiguities must be given, together with the objects themselves, as something which cannot be reduced to something else and which requires no reduction. . . . The subject matter of mathematics is, in accordance with this theory, *the concrete symbols themselves whose structure is immediately clear and recognizable*.

(Hilbert 1925: 191–92, my italics)

Here is the core idea of formalism: mathematics is about symbols. But Hilbert's Kantian idea is now to study these symbols mathematically. What kind of mathematics is used? Not the questionable infinite stuff, but rather finite, meaningful mathematics, intimately linked to the perception of concrete objects, to the perception of the (finitely many) concrete *symbols* of classical mathematics itself.

Hilbert's Programme

The study of mathematics can itself be mathematical. But the mathematics used must be true, meaningful, finite mathematics. If we are worried about the consistency of, say, transfinite set theory, we can hardly use that very set theory to check its own consistency. But we might be able to establish its consistency using the utterly reliable techniques of finite mathematics. How would this be done? Not by showing that there really are infinite sets which are correctly described by the theory. Rather, we should focus on the concrete, perceivable symbols of set theory. Since a proof, according to Hilbert, is just a sequence of symbols manipulated according to the rules, we need to show that there is no sequence of symbols that results in, say, the expression '$\omega \neq \omega$' or '$0 = 1$'. (Any absurdity will do. Since an inconsistent theory implies everything, showing that it doesn't imply some particular absurdity will in effect show that it implies none.)

There are actually a number of different ways to show consistency. In the nineteenth century non-Euclidean geometries were shown to be consistent by constructing Euclidean models of them. These are known as 'relative consistency proofs', since the consistency of non-Euclidean geometry rests on the *assumed* consistency of Euclidean geometry. Hilbert himself in his work on the foundations of geometry (Hilbert 1899) gave a further consistency proof of Euclidean geometry, again a relative consistency proof, using the real numbers as the model. But what about the reals themselves? And can we do better than merely providing relative consistency proofs? What about an absolute proof of consistency?

Sometimes a theory can be shown to be consistent simply by exhibiting a concrete model. The proof of consistency comes through visual inspection of the model. This could count as an absolute proof. Here is a simple example, which I'll call the theory *T*:

Primitives of *T*: some objects: *ponk, lonk*, and a relation: *zonks*.
Axioms of *T*: (1) For any two lonks, at most one ponk zonks both.
(2) For any two ponks, exactly one lonk zonks both.
(3) There are at least three ponks which zonk each lonk.

Of course, this sounds like meaningless gibberish – indeed, it is. But we might want to know whether it is *consistent* gibberish or not. Perhaps we can derive something like: 'There is a lonk with exactly one ponk which zonks it' which contradicts the third axiom. The following concrete model (Figure 5.1) shows that *T* is actually a consistent theory.

The model is sometimes called the seven-point geometry. There are exactly seven points (the indicated vertices) and exactly seven lines (the circle is taken to be a line). We interpret 'ponk' to be one of the seven points, 'lonk' to be one of the seven lines, and 'zonks' to be the relation, *lies on* or *connects*. So understood, the axioms are true. To see this, examine each pair of lines and notice that they have exactly one point in common. This shows that the first axiom is true under this interpretation. Similar considerations show the other axioms true. And since logical inference preserves truth, no falsehood can be derived from these axioms; hence no contradiction can arise. *T* is a consistent theory.

In proving the consistency of *T* we have focused on the concretely visual; the mathematics involved is strictly finite (i.e. counting lines, etc.). The *meaning* of terms such as 'ponk' (aside from the interpretation) play no role whatsoever in establishing the consistency of *T*.

No serious theory, of course, will be so easily tackled. No simple concrete model of, say, set theory or complex analysis showing at a glance its consistency is forthcoming. Real ingenuity will be required. But even with all the ingenuity in the world, as we shall see, it still won't work, not even for lowly arithmetic.

Hilbert's programme called for a number of things. First, all existing theories would have to be formalized. Classical mathematics (as found in typical texts

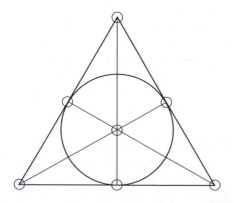

Figure 5.1 The seven-point geometry as a model of *T*

and research papers) is rather informal, a mix of symbols and natural language. It would all have to be recast into precise and exact symbolic notation. Much of this sort of work, Hilbert was happy to note, had already been carried out, for example, in Whitehead and Russell's *Principia Mathematica* (1910). Second, the notion of 'finite method' would have to be made perfectly precise. An intuitive distinction between finite and non-finite techniques was available, but a much sharper notion was needed. Finally, these finite methods would be applied to the formalized versions of classical mathematics to uncover some of their properties and, above all, to show them consistent.

Hilbert's programme was brilliantly conceived and (partially) executed. The original ideas for the plan of attack and the extent of the achievements are stunning. Some of the most brilliant mathematicians/philosophers of the century worked on the programme, including: Ackermann, Bernays, von Neumann, and many others, besides the amazing Hilbert himself. It is now generally thought that Gödel's incompleteness theorems dashed all hope for the programme. This will be explained below. But even if the main programme failed, the by-products boggle the mind: axiomatics, model theory, recursive function theory, theory of algorithms and computation, and much, much more. The brilliance of Hilbert and his co-workers in the foundations and philosophy of mathematics should not be diminished – this cluster of accomplishments is one of the intellectual highlights of the twentieth century.

Small Problems

Needless to say, there have been lots of objections to Hilbert's formalism, big and small. The big one, which is utterly devastating, stems from Gödel's famous results. I'll come to that in a moment. First, I'll take up some of the smaller problems, concerning complex finitistic reasoning.

Hilbert associates trustworthy reasoning with the finite. But, clearly our grip on a finite entity lessens as the entity become larger and more complex. We can multiply two small numbers together and be confident of the answer. But what about the product of two finite numbers each over a billion digits long. It certainly does not correspond to any object of perception in the Kantian sense. The chances of making a computational error are considerable. I, for one, would be much happier betting on the truth of a transfinite proposition such as $2^\omega > \omega$, than on an enormously complex finite example. Certainty cannot be simply identified with the finite; at best it can be linked to the rather small. But if we confine ourselves to this, we won't have anything close to the classical mathematics we want and need.

Moreover, as Hilbert correctly noted, the universe is finite. So for some large but still finite number, there cannot be any appropriate perceptual experience (i.e. there aren't enough ||||||...). Thus, even large finite numbers will have to be classified along with transfinite entities as fictions or ideal elements.

On the other hand, perhaps these objections needn't be seen as problems in principle with Hilbert's programme. It might turn out that the consistency proofs that are required to justify classical mathematics all happen to be short and manageable. Of course, this is a moot point, since there are no such proofs – long or short.

Gödel's Theorem

Gödel's theorem, which shows the incompleteness of any attempt to system-atize arithmetic, is perhaps the single most famous and important result in logic in this or any century. It is also one of the most difficult to understand. Recently, however, George Boolos has found a much simpler proof of the incompleteness theorem. The proofs of Gödel and Boolos are each modelled on famous, but dif-ferent, paradoxes. Gödel makes use of the Liar Paradox (*What I am now saying is false* – a paradox because if true then it's false, and if false then it's true.) He formulates a sentence of arithmetic which says something like 'I am not prov-able', and sure enough it turns out to be unprovable, so it would seem to be a true but unprovable sentence of arithmetic, making the system of axioms incomplete. Boolos's proof is modelled on the Berry Paradox (*The least integer not namable in fewer than nineteen syllables* – a paradox because I just named that integer in eighteen syllables.) Even though his proof is much simpler than Gödel's, it still requires a fair bit of work.

A Formal System Of Arithmetic

The statements in the language, *L*, of formal arithmetic are built out of the fol-lowing symbols:

+	addition
×	multiplication
0	zero
s	successor (i.e. plus one)
=	equals
~	not
∨	or
&	and
→	if . . . then . . .
↔	if and only if
(	left parenthesis

)	right parenthesis
$\forall$	all
$\exists$	some
$x, x', x'', \ldots$	variables (infinitely many, but only two symbols: x and $'$; for ease of reading we will often use y, z, and w in place of the primed variables, x', x'', x'''.)

The symbols 1, 2, 3, ... are not part of this formal language, but they have counterparts which are: s0, ss0, sss0, ... The variables x, x', x'', ... have these numbers as their values.

Truth

The intended interpretation of this formal arithmetic is just the ordinary natural numbers, and statements of the formal system are true or false just in case their intended interpretations are true or false. For example:

- '$\forall x \exists y (y = sx)$' is true (in the intended interpretation) because each number has a successor, namely, the next bigger number.
- '$\forall x \exists y (x = sy)$' is false because 0 is not a successor of any number.
- 'ss0 + sss0 = sssss0' is true since $2 + 3 = 5$.
- 'ss0 + sss0 = sssssss0' is false because $2 + 3 \neq 8$
- '$\forall x \exists y (x = (y + y) \vee x = s(y + y))$' is true since every number is odd or even.

Axiomatization

From this point there are a number of ways we could proceed. One way is to set up an algorithm. This would be a computational procedure (performed by a computer, for example) which generates a certain output. A correct algorithm for formal arithmetic would be an algorithm that generated all the truths of arithmetic and none of the falsehoods. A second way to proceed would be to specify a set of axioms (such as the Peano axioms) and then consider the set of logical consequences of those axioms. An axiomatization of formal arithmetic is consistent and complete if and only if it implies all the truths of arithmetic and none of the falsehoods.

Here is a particular axiomatization, *PA*, based on the Peano axioms:

(1) $\forall x \forall y (sx = sy \rightarrow x = y)$
(2) $\forall x \sim (0 = sx)$
(3) $\forall x \exists y (\sim (x = 0) \rightarrow x = sy)$
(4) $\forall x (x + 0 = x)$
(5) $\forall x \forall y (x + sy = s(x + y))$

(6) $\forall x(x \times 0 = 0)$
(7) $\forall x \forall y(x \times sy = (x \times y) + x)$
(8)   For any sentence S(x), (S(0) & $\forall x$(S(x) → S(sx))) → $\forall x$(S(x))

The last axiom expresses the principle of mathematical induction. It is known as an axiom *schema*, since it is really infinitely many axioms (one for each different sentence S). From these axioms one could prove such things as ss0 + ss0 = ssss0 (i.e. 2 + 2 = 4) or any of the examples cited above. Here's a simple illustration.

⊢ ss0 + ss0 = ssss0:

(1) $\forall x \forall y(x + sy = s(x + y))$ Axiom (5)
(2) $\forall y(ss0 + sy = s(ss0 + y))$ (1) Universal instantiation
(3)   ss0 + s0 = s(ss0 + 0)   (2) Universal instantiation
(4) $\forall x(x + 0 = x)$ Axiom (4)
(5)   ss0 + 0 = ss0   (4) Universal instantiation
(6)   ss0 + s0 = s(ss0)   (3), (5), (6) Equality
(7)   ss0 + ss0 = s(s(ss0))   (2), (6) Equality
(8)   ss0 + ss0 = ssss0   (7) Brackets removed.

Care to test your strength? As an exercise show that the following can be derived using the above axioms: (a) ⊢ $\forall x(\sim(x = sx))$, (b) ⊢ $\forall x \forall y(x + y = y + x)$, (c) ⊢ ssx + sssx = sssssx. The answer to (a) is given on p. 222.[4]

The interesting question, however, is whether this axiomatization or any other can capture all the truths of arithmetic and none of the falsehoods. The answer is No. This was first shown by Gödel (1931). His proof or variants of it can be found in almost any text on the subject.[5] The proof below is due to Boolos (1989).

The Boolos Proof

Theorem: No consistent algorithm/axiomatization of formal arithmetic is complete.

The proof works as follows: we will assume there is an algorithm or axiomatization, **A**, which generates truths of arithmetic, but no falsehoods; we will show that no matter what **A** is, there will always be some truth of arithmetic that **A** fails to generate. Thus, **A** is not complete. The system *PA* is a special case of **A**, so *PA*, in particular, is incomplete.

Let [n] be the expression consisting of 0 with n successor symbols in front. Thus, [5] is the expression sssss0 (and so [5] stands for 5).

Definition: The formula F(x) *names* the natural number n if the following statement is generated by **A**: $\forall x(F(x) \leftrightarrow x = [n])$.

Here's an example: *PA* generates $\forall x(x + x = \text{ssss}0 \leftrightarrow x = \text{ss}0)$ that is, proves it as a theorem; so the formula $x + x = \text{ssss}0$ names the number 2.

The formula F(x) has some crucial properties. First, it can name only one natural number (though other formulae might also name the same number). Second, for each number i, there are only a finite number of formulae that contain i symbols. This means that there are only finitely many numbers named by formulae containing i symbols. Third (which follows from the second point), for any number j, there are only finitely many numbers named by formulae containing fewer than j symbols. Fourth (which follows from the third), there must be a least such number.

For those who like to prove things, have a go at the first remark, i.e. prove that no formula can name two different numbers. Answer on p. 222.[6] If you're still keen, prove the first part of the second remark, i.e. for each number i, there are only a finite number of formulae containing i symbols. Hint: there are only 16 primitive symbols in the formal system. Answer on p. 222.[7]

For the next stage in the proof we need the idea:

> *x is named by a formula containing z symbols*

So we posit a formula C(x,z) in the language L which says this. The example above illustrates the point: the formula $x + x = \text{ssss}0$ names the number 2 and it has 9 symbols; thus, C(2,9). The existence of such a formula C(x,z) may seem fairly plausible, but actually showing that for any algorithm/axiomatization, **A**, such a C(x,z) exists is rather difficult and tedious. Among other things we would have to show that a process of coding, such as Gödel numbering, leads to certain properties which allow statements about numbers to be represented in the system of formal arithmetic. We shall take this to be established, and hence that there will always be such a formula C(x,z) expressed in the formal language L.

The next concept we need is:

> *x is a number which is named by a formula which has fewer than y symbols*

We can express this in L by B(x,y) which is defined to be the formula $\exists z(z < y \,\&\, C(x,z))$. (Note that the new symbol $<$ can be defined in L as follows: $x < y \leftrightarrow \exists z\,(z \neq 0 \,\&\, x + z = y)$.)

We need one more concept to be able to specify an analogue of the Berry Paradox:

> *x is the least number that is not named by any formula containing fewer than y symbols*

To formalize this in L, let $A(x,y)$ be the formula $\sim B(x,y)$ & $\forall w(w < x \rightarrow B(w,y))$. What $A(x,y)$ says is precisely what we want.

We'll let k be the number of symbols in $A(x,y)$; note that $k > 3$. (We'll need this fact later.)

Next we'll let $F(x)$ be the formula $\exists y(y = ([10] \times [k])$ & $A(x,y))$. $F(x)$ says that x is the least number not named by any formula containing fewer than $10k$ symbols. Does $F(x)$ pick out a number? Obviously, Yes. $F(x)$ specifies some number n which is indeed the least number not named by any formula containing fewer than $10k$ symbols.

Here's an easy quiz to keep you on your toes. How many symbols does [10] contain? More generally, how many symbols does [n] contain? Answer on p. 222.[8]

We can easily calculate how many symbols $F(x)$ contains:

(1) [10] contains   11 symbols
(2) [k] $k + 1$
(3) $A(x,y)$ k
(4) others   12 (i.e. $\exists, x,',(, x,',(, = ,\times,),\&,)$)
(5) TOTAL $2k + 24$

Notice that $2k + 24 < 10k$ for any allowable value of k, since it was noted above that $k > 3$. Also above it was shown that for any number j, there is a least number not named by any formula containing fewer than j symbols. Set $j = 10k$ where n is the least such number for this j. Thus, n is not named by the formula $F(x)$, since, if it were, it would be named in fewer than $10k$ symbols. This means that **A** (if it is consistent) does not generate $\forall x(F(x) \leftrightarrow x = [n])$. However, this is a true statement since, as mentioned above, n is indeed the least number not named by any formula containing fewer than $10k$ symbols. Therefore, **A** is incomplete.

Gödel's Second Theorem

The famous incompleteness result actually consists of two parts. One is the incompleteness result just presented, the other is the theorem which says that there can be no finitistic proof of the consistency of a system of arithmetic within that system. This second theorem is actually the more devastating of the two for Hilbert's programme. I will follow (or at least adapt) a recent, clever version of the proof given, yet again, by George Boolos (1994).

We can prove all sorts of things in the system **A**, such as that $5 + 7 = 12$; and we can prove that we can prove it, and so on. And we can prove that other things are not true, such as that $5 + 7 \neq 13$. Moreover, we can prove that we can prove that. The consistency worry is that we might be able to prove too much.

If **A** is *not* consistent, then we can prove everything, including that $0 = 1$ (i.e. $\vdash 0 = s0$). If we could prove that there is no proof of this, then we'd know that **A** is consistent. Of course, the proof will have to pass Hilbert's finitistic standards, otherwise it might be question-begging.

Amazingly, it turns out that we cannot prove **A** (or any other such system) consistent. This is Gödel's second theorem and it delivered a knockout blow to Hilbert's hopes.

I'll use the notation $\exists\wp(p)$ to mean there is a proof of p (or, more exactly, there is an x such that x is a proof of [p], where [p] is the appropriate representation of p in **A**, say by a Gödel number). Using this notation we can now define the *consistency* of **A** very simply as $\sim\exists\wp(0 = s0)$, and the *proof of consistency* will be $\exists\wp\sim\exists\wp(0 = s0)$. Gödel's second theorem can now be easily stated.

> *Theorem*: If **A** (or any other system of arithmetic) is consistent, then there is no proof in **A** of the consistency of **A**; that is, if $\sim\exists\wp(0 = s0)$ then $\sim\exists\wp\sim\exists\wp(0 = s0)$.

Before getting to the proof, we will adopt the following special rules (known as the Hilbert–Bernays–Löb derivability conditions), as well as the usual rules of logic:

> Rule I   if $\vdash$ p then $\vdash \exists\wp$p (parenthesis around p dropped when obvious)
> Rule II $\vdash (\exists\wp(p \rightarrow q) \rightarrow (\exists\wp p \rightarrow \exists\wp q)$
> Rule III $\vdash (\exists\wp p \rightarrow \exists\wp\exists\wp p)$

These rules have the consequence

> Rule IV   if $\vdash (p \rightarrow q)$ then $\vdash (\exists\wp p \rightarrow \exists\wp q)$

Try your hand at deriving the fourth rule from the other three. Answer on p. 222.[9] Now we can get on with the proof.

Proof of Gödel's second theorem: In the theorem above showing incompleteness we established the existence of a sentence of **A** (call it g in honour of Gödel) that is equivalent to its own unprovability; this is our starting point here:

(1)	$\vdash g \leftrightarrow \sim\exists\wp g$	from the first incompleteness theorem
(2)	$\vdash g \rightarrow \sim\exists\wp g$	(1)
(3)	$\vdash \exists\wp g \rightarrow \exists\wp\sim\exists\wp g$	(2), Rule IV
(4)	$\vdash \exists\wp g \rightarrow \exists\wp\exists\wp g$	Rule III
(5)	$\vdash \sim\exists\wp g \rightarrow (\exists\wp g \rightarrow (0 = s0))$	tautology
(6)	$\vdash \exists\wp\sim\exists\wp g \rightarrow \exists\wp(\exists\wp g \rightarrow (0 = s0))$	(5), Rule IV
(7)	$\vdash \exists\wp(\exists\wp g \rightarrow (0 = s0)) \rightarrow (\exists\wp\exists\wp g \rightarrow \exists\wp(0 = s0))$	Rule II
(8)	$\vdash \exists\wp g \rightarrow \exists\wp(0 = s0)$	(3), (6), (7), (4)

$$(9) \quad \vdash \sim\exists\wp(0 = s0) \to g \qquad\qquad (1), (8)$$
$$(10) \quad \vdash \exists\wp\sim\exists\wp(0 = s0) \to \exists\wp g \qquad (9), \text{Rule IV}$$
$$(11) \quad \vdash \sim\exists\wp(0 = s0) \to \sim\exists\wp\sim\exists\wp(0 = s0) \qquad (8), (10)$$

The last line is our theorem, namely, if **A** is consistent then there is no proof of consistency in **A**. To see this clearly, suppose we could prove consistency, that is prove $\vdash \sim\exists\wp(0 = s0)$. Then from (11), by *modus ponens*, we would have $\vdash \sim\exists\wp\sim\exists\wp(0 = s0)$. But by Rule II we would also have $\vdash \exists\wp\sim\exists\wp(0 = s0)$. A contradiction. Thus, assuming consistency of **A**, there can be no proof of it.

The Upshot for Hilbert's Programme

The consequence of Gödel's two theorems are manifestly clear and generally acknowledged. First of all, the formalist hope of identifying *truth* with *provability* flounders on the first incompleteness theorem, since in any consistent theory there will always be true but unprovable sentences. Since this applies to any theory strong enough to contain arithmetic, it applies in effect to all of classical mathematics. Second, the impossibility of a consistency proof is even more destructive. Hilbert's hopes of giving a (finitistically acceptable) proof of the consistency of classical mathematics are completely dashed.

I should note in passing that Gödel's original proof of the incompleteness theorem is constructive, and therefore intuitionistically valid. (More on these notions in Chapter 8.) Boolos's proof is not constructive. Those who demand an intuitionistically valid proof in order to consider a theorem legitimate will have to stick with the original, constructive version of the proof.[10] In so far as Hilbert's Programme was undermined by Gödel's original proof, it is still hurt by this one. Non-constructive methods *cannot* be used in *support* of Hilbert's programme, since that would be question-begging; but *negative* results, such as this one, can use non-constructive methods in a non-circular way. (As a final exercise for this chapter explain and justify this last remark.)

The Aftermath

It is almost universally agreed that Gödel's results destroyed Hilbert's programme, but there are hold-outs. Abraham Robinson (one of the creators of non-standard analysis) claimed that in spite of the incompleteness theorems we should be formalists anyway (Robinson 1964). More recently Detlefsen (1986, 1992a) has claimed that much of Hilbert is left undamaged by Gödel. Robinson's claims are stunning, given all that he concedes. Detlefsen more carefully argues his case. However, I won't pursue these attempts to keep the Hilbert programme alive.

More controversial have been the possible implications of Gödel's results for the mind and for Platonism, in particular that the human mind cannot be a machine and that Platonism must be correct. Gödel himself suggested this, and Lucas, in a famous article argued explicitly that that mechanism is wrong (Lucas 1961). Most recently Roger Penrose (1989, 1994) has argued at great length for both theses, claiming that the Gödel results show that the whole programme of artificial intelligence is wrong, that creative mathematicians do not think in a mechanistic way, but that they often have a kind of insight into the Platonic realm which exists independently from us.

I won't pursue the consequences of Gödel's results other than their impact on Hilbert's programme. I mention the work of Penrose and others only to inform the reader who might not be aware of its existence and relevance to the philosophy of mathematics. But one brief comment is in order before moving on to other things. I am largely persuaded of Penrose's conclusions concerning mathematics. The real question concerns his argument for those conclusions. It is bold and subtle. It may be wrong, but it is plausible. And I suspect that it certainly deserves a better hearing than suggested by Hilary Putnam who 'regards its appearance as a sad episode in our current intellectual life' (1994: 1). It's a common failing of philosophers to ignore people's claims based on special experiences. Ethics, for example, is needlessly impoverished by ignoring, say, the daily life of mothers raising children under difficult circumstances; it would be much improved by paying closer attention. Roger Penrose – modesty forbids him saying it of himself – has enjoyed some of the most profound mathematical experiences of recent times. If he has nothing more than a mere hunch that he is glimpsing into the Platonic realm, that in itself is something for us all to ponder.

Further Reading

Hilbert's main works can be found in translation in various anthologies, such as Benacerraf and Putnam (eds.), *Readings in the Philosophy of Mathematics*; Ewald (ed.), *From Kant to Hilbert*; Mancosu (ed.), *From Brouwer to Hilbert*. Reid's biography, *Hilbert*, is interesting and informative on his whole life as well as his work on foundations. The chapter on formalism in Shapiro's *Thinking About Mathematics* is very clear. Various essays by Hallett and by Seig (see bibliography) are exemplary historical studies. Web's *Mechanism, Mentalism, and Mathematics* covers lots of relevant material. Some work on formalism can be found in various discussions of Godel's theorem. Formalism is largely rejected today, but Detlefsen, *Hilbert's Program, an Essay on Mathematical Instrumentalism*, tries to rescue much of Hilbert's programme.

CHAPTER 6
Knots and Notation

Formalists have glummed onto something fundamental. Their great insight is in noticing just how important notation can be. Of course, the identification of mathematics with its symbols – the very essence of formalism – is a mistake. But the realization that notation means more to mathematics than the corresponding special symbols of other sciences means to them, seems to me profoundly correct.

In one regard, mathematics is like poetry. Every discipline from auto mechanics to zoology uses its own system of representation, a specialized vocabulary and system of symbols that helps to convey the old facts and new speculations in the field. But poets go beyond this. They fix on the form of representation itself, then exploit it in highly creative and beautiful ways. Perhaps every discipline does this to some extent, but only mathematics matches poetry in tying innovation to notation.[1]

To remind ourselves of how clever poetry can be in this regard, let's consider a few lines from Pope's ingenious *Essay on Criticism*, in which he exemplifies the interplay of form and content while explaining it.

'Tis not enough no harshness gives offence,
The sound must seem an echo to the sense.
Soft is the strain when Zephyr gently blows,
And the smooth stream in smoother numbers flows;
But when loud surges lash the sounding shore,
The hoarse, rough verse should like a torrent roar.
When Ajax strives, some rock's vast weight to throw,
The line too labours, and the words move slow;
No so, when swift Camilla scours the plain,
Flies o'er th'unbending corn, and skims along the main.

Not all notational cleverness leads to good results. Milton sometimes relied on 'eye rhymes', in which spellings fit, not sounds, for example, in *On the University Carrier*:

He's here stuck in a slough, and overthrown.
'Twas such a shifter, that if truth were known
Death was half glad when he had got him down . . .

The rule: *seen and not heard*, is bad for children, terrible for rhymes. And this latter example shows that in poetry the principal notation (if I may speak that way), is the sound, not the written word; ideas and feelings (the content) are conveyed by means of sounds (the form); the written symbols are representations of the sounds. Milton departed from this; others have, too; for example, e.e. cummings often made a poem take a particular physical shape on the page. None of this has ever been very successful, although Chinese poetry, because the written form of Chinese is character-based and lends itself to beautiful and suggestive calligraphy, often plays very successfully on the clever mixing of content with both sound and writing.

Whatever the ins and outs of poetry, one thing is clear: the manner of expression – notation – is fundamental. It is the same with mathematics – not in the aesthetic sense that the beauty of mathematics is tied up with how it is expressed – but in the sense that mathematical truths are revealed, exploited and developed by various notational innovations.

Perhaps the greatest notational invention of all time is the Arabic numerals. Though these numerals name infinitely many distinct numbers, we can easily get the hang of this notational system and figure out the name of any particular number. I can't remember the names of all my students, but I have no trouble with the name of any natural number. Line up a few strokes: | | | | | . . . and any child can rattle off the names of the associated numbers with ease. Many of the key properties of numbers are built right into their names. For example, we need only look at the names '123' and '23' to realize the first of these refers to a bigger number than the second. And a mere glance at '6137594' tells us that it names a composite number. By contrast, we may study the names 'proton' and 'electron', with no hope of ever learning that protons are heaver than electrons.

This may seem a trivial point, but it is only because we are so used to the phenomena. Often it is noted that the Arabic notation is vastly superior to the Roman. True, but even the oft-ridiculed Roman system of numerals is ingenious when compared to the possibility that each number is given its own distinct name. Instead of 1, 2, 3, 4, . . . or I, II, III, IV, . . . try working with numbers which have names like George, Mary, Bill, Ann, ... We'd have theorems such as: George + Mary = Bill, but imagine trying to prove them. The crucial feature that is built into both Roman and Arabic numerals is their recursiveness – there is an algorithm for employing them.[2] The recursive properties of the natural numbers are mirrored in the notation.

Clever notations do not come out of the blue. Developing a notation and learning about the objects named often go hand-in-hand. But one thing is sure: we could not have *first* invented a system like Arabic numerals and then *later* discovered that the natural numbers form a recursive set. In some sense, this must have already been realized, however dimly.

When we reflect on an example like the recursive Arabic numerals, we can readily see the powerful attraction of formalism as a philosophy of mathematics. It is not the formalist claims about mathematics being about ink marks on paper – a silly view inspired by a nominalistic hostility to abstract entities. Rather the source of the attraction of formalism stems from the evident power of notations themselves. The crucial thing about the natural numbers is their recursiveness. But this key property is also in the notation itself. So why bother with numbers, the would-be formalist wonders, when the *numerals* themselves possess all the characteristics we care about?

There is no denying the attractiveness of such a view. Ultimately, though, it is untenable. Quite aside from special difficulties such as those which arise with Gödel's theorem, the problem with this view is that it puts the cart before the horse. As I mentioned already, the recursive Arabic numerals could only be invented *after* recognizing the recursiveness of the natural numbers themselves. There is a sense in which the notation of number theory is an application of number theory. There is no understanding of that notation without a prior understanding of the objects named. This is certainly not true of all names, but it is often the case, as with an ingenious notation like the Arabic numerals. This is the mathematical equivalent of poetry's onomatopoeia.

Knots

Knot theory provides an exceedingly rich example of how different notations – different forms of representation – bring out different aspects of the common subject matter. We are naturally familiar with knots from daily life. In mathematics a *knot* is defined as a closed, non-intersecting curve in space. They can be twisted and tangled, and generally transformed in various ways, and still remain the same knot, so long as they are not cut. Properties which hold through such deformations are known as *invariants*. The fundamental problems of knot theory involve finding ways to distinguish and classify different knots. The theory got its impetus long ago from physics. Lord Kelvin in the 1880s conjectured that atoms are knotted vortices in the aether, with different chemical properties due to different knots. This leads quite naturally to a classification of knots, undertaken by Kelvin's longtime co-worker P.G. Tait. Interest in knots declined when this account of atoms was rejected and has remained at a low level for most of the twentieth century. Recent years, however, have seen an enormous growth of interest due to some stunning results and amazing connections to physics.

Two knots are *equivalent* when one curve can be deformed into the same shape as the other without either cutting the curve or allowing a self-intersection during the transformation. We might represent a knot with a particular piece of string joined together at the ends, or an electrical extension cord with the ends plugged in. These are knots in space. A more common representation is a drawing on the page, known as a *projection*. Figure 6.1 gives some examples.

The allowable moves in any transformation (of a projection of a knot) are known as *Reidemeister moves*. There are three of them. The first of these allows us to put a twist in the knot, or to remove one, while the rest of the knot remains unchanged. The other two moves will be similarly obvious from Figure 6.2.

Reidemeister proved that for any two projections of the same knot, there is a sequence of Reidemeister moves which will transform one projection into the other. This is one way of showing them to be equivalent. In Figure 6.3 we have a tangled mess transformed by a sequence of Reidemeister moves into the

The unknot, the trefoil, and a tangled mess which is equivalent to the unknot

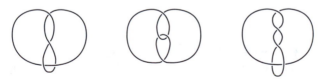

Three different projections of the figure-eight knot

Figure 6.1

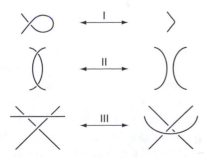

Figure 6.2 The three Reidemeister moves

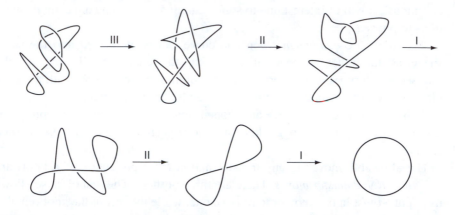

Figure 6.3 A sequence of Reidemeister moves resulting in the unknot

unknot, thereby showing them equivalent. There is, however, no known algorithm for applying Reidemeister moves; it's just trial and error. It's not even
known whether there are any upper bounds. Is there, for example, a number k
such that in transforming one knot with n crossings into another, all intermediate knots have fewer than $n + k$ crossings? This is one of a great many
open problems in knot theory.

The Dowker Notation

We begin by choosing an orientation; that is, we put an arrowhead on some
strand of the knot. Then we pick a crossing and label it 1. Next, move along the
understrand in the direction of the arrow until we reach the next crossing, which
should be given the label 2. If an even number assigned to any crossing is an
overstrand it should be considered a positive number; but if it is an understrand,
then it is negative. The odd numbers are positive, regardless. We continue on in
this way until we are back where we started. Each crossing will have two numbers associated with it, one even and one odd. (As an exercise, think about it for
a moment and you will see why. Answer on p. 222.[3])

The knot in Figure 6.4 can then be described in the Dowker notation as a set
of ordered pairs: $\{\langle 1,4 \rangle, \langle 3,-6 \rangle, \langle 5,10 \rangle, \langle 7,-2 \rangle, \langle 9,8 \rangle\}$. Since the first member
of each pair is odd and contains no information about $+$ or $-$, we can express
this quite economically as a sequence of even numbers: $\langle 4,-6,10,-2,8 \rangle$. The
process is obviously reversible: given a sequence, we can construct the knot
projection. All the information we need is in the Dowker sequence: it tells us
how many crossings there are, how they are connected, and which strands are
over or under.

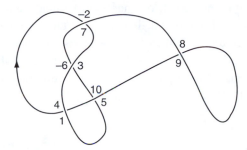

Figure 6.4 A knot labelled with the Dowker notation

I said the process is reversible, but this is not completely true. The Dowker notation does not completely determine what are called composite knots and mirror images. (I won't go into these except to say that composite knots are composed out of prime knots – think of them as constructed out of two or more simpler knots. And the mirror image of a knot is just what you would think it to be.)

The Dowker notation, because it uses a simple sequence of numbers, lends itself to computer programming rather nicely. It has been used for the tabulation of all prime knots up to thirteen crossings, of which there are 9988. No one knows how many there are at fourteen crossings, but finding a suitable notation for tackling the problem is half the battle. Another interesting feature of the Dowker notation is that it readily shows some of the crossings to be trivial in the light of Reidemiester moves. The crossing <8,9>, for example, can be eliminated by a Reidemeister I move. This will always be the case when a crossing is labelled by two successive numbers. Type II moves are also captured by this notation.

The Conway Notation

A *tangle* is any region of the projection plane which is surrounded by a circle in such a way that the knot crosses the circle exactly four times. And, as you might expect, two tangles are equivalent when one can be transformed into the other by a series of Reidemeister moves (with the condition that the four strings leading out of the circle remain fixed and that the tangle remains wholly within the circle).

Figure 6.5 A tangle

Here in Figure 6.6 are some special cases. A pair of uncrossed vertical lines is called the ∞ tangle; a pair of uncrossed horizontal lines is the 0 tangle; a pair of lines crossed three times is the 3 tangle. If they had been twisted the other way (i.e. a right-handed twist instead of a left-handed twist), it would be the -3 tangle. Looking from left to right, if the slope of the overline is up, then the twist is left-handed and positive; if the overline is headed down, then it is a right-hand twist and negative.

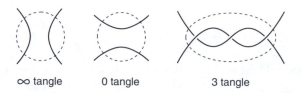

∞ tangle   0 tangle   3 tangle

Figure 6.6 Tangles with Conway notation

We can use this notation to characterize more complex tangles. For instance, in Figure 6.7 we start with a 3 tangle (Figure 6.7(a)). After a mirror reflection, we have Figure 6.7(b). We then twist the two right strands. The resulting tangle (Figure 6.7(c)) is denoted 3 2. We then do another mirror reflection (Figure 6.7(d)). (Notice that we are always working towards the right.) This is followed by twisting the two right strands, this time giving them a negative twist four times (Figure 6.7(e)). The result expressed in the Conway notation is 3 2 -4.

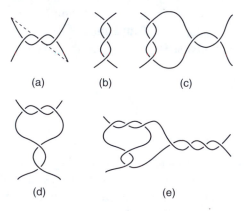

(a)   (b)   (c)

(d)   (e)

Figure 6.7 The construction of a 3 2 -4 tangle

We can use these numbers to construct a continued fraction. Working this time from right to left, the continued fraction associated with 3 2 -4 is

$$-4 + \cfrac{1}{2 + \cfrac{1}{3}}$$

Simplified, it is equal to $-25/7$.

Amazingly, we have the following theorem: *Two tangles are equivalent if their associated continued fractions are equal.* What starts out as a mere labelling device quickly turns into a powerful computational tool for investigating the properties of knots. Imagine being able to determine Joe's height, weight, and other characteristics just by studying his name.

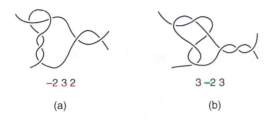

$$-2\ 3\ 2 \qquad\qquad 3\ -2\ 3$$

(a) \qquad\qquad (b)

Figure 6.8 Conway notation applied to equivalent tangles

For example, Figures 6.8(a) and 6.8(b) are denoted $-2\ 3\ 2$ and $3\ -2\ 3$, respectively. Their continued fractions are:

$$2 + \cfrac{1}{3 + \cfrac{1}{-2}} = 3 + \cfrac{1}{-2 + \cfrac{1}{3}} = \frac{12}{5}$$

To convince yourself of this theorem (at least for the case at hand), transform one tangle into the other by a sequence of Reidemeister moves.

Polynomials

Another exceedingly interesting representation of knots is by means of polynomials. This particular notation was discovered by Alexander early in this century; in the past decade several other forms have been discovered leading to some quite spectacular results. I'll briefly describe one of the simpler of these, the bracket polynomial and, unfortunately, ignore others such as the Jones polynomial (which started the new wave in 1984), the HOMFLY polynomial (an acronym derived from several simultaneous discoverers), and the amazing connections to physics which have shed even more light on knot theory (see Jones 1990 and Witten 1989).

The first rule says that the polynomial for the unknot is just the number 1.

Rule 1: $\langle \bigcirc \rangle = 1$

The next rule says that given a crossing, split it into the sum of two projections, each with one fewer crossing, and a coefficient, so far undetermined.

Rule 2: $\langle X \rangle = A \langle)(\rangle + A^{-1} \langle \asymp \rangle$

Notice the polynomial to be constructed is a Laurent polynomial, which has both positive and negative exponents. The third rule extends the discussion to links as well as knots. A *link* is a set of knots tangled together.

Rule 3: $\langle L \cup \bigcirc \rangle = (-A^2 - A^{-2})\langle L \rangle$

We can now easily calculate the polynomial for the unlink:

$$\langle \bigcirc \cup \bigcirc \rangle = (-A^2 - A^{-2})\langle \bigcirc \rangle$$
$$= -A^2 - A^{-2} \text{ (since } \langle \bigcirc \rangle = 1)$$

The Hopf link is slightly more challenging.

$$\langle \infty \rangle = A \langle \text{CO} \rangle + A^{-1}\langle \text{CO} \rangle$$
$$= A(A \langle \text{CO} \rangle + A^{-1}\langle \text{CO} \rangle)) + A^{-1}(A \langle \text{CO} \rangle + A^{-1}\langle \bigcirc\bigcirc \rangle)$$
$$= A(A(-(A^2 + A^{-2})) + A^{-1}(1)) + A^{-1}(A(1) + A^{-1}(-(A^2 + A^{-2})))$$
$$= -A^4 - A^{-4}$$

Why are these polynomials interesting or important? The key is in realizing that they are invariant under Reidemeister moves. (There is an important quali-fication to be made with respect to type I moves, but I won't go into that here.) When we calculate a type II move, for instance, we get:

$$\langle X \rangle = A \langle \asymp \rangle + A^{-1} \langle)\neg (\rangle$$
$$= A \langle \asymp \rangle + A^{-1} \langle)\!-\!(\rangle$$
$$= \langle X \rangle$$

The Whitehead link

Borromean rings

The Unlink of
two components

The Hopf link

Figure 6.9 Examples of links

Since equivalent knots can be transformed into one another by Reidemeister moves, they must also, in consequence, have the same associated polynomial. On the other hand, knots with different polynomials must be distinct knots.

Note the curious way these calculations have been carried out. Though obvious, I want to stress it anyway, since it's so philosophically important. We wrote down things that look like equations. Indeed they were. But pictures occurred in these equations. This notation, which is widely used and which is extremely efficient in calculating polynomials, is certainly not like other verbal/symbolic statements; it is more like hieroglyphics, a form of picture writing. As calculations, these are picture-proofs.

Creation or Revelation?

Mathematicians and historians often mention the importance of a good notation. But serious analyses have so far been scarce. Ken Manders, however, has recently put his finger on several key points and developed some interesting claims.

'Mathematical practices', he says, 'pursue their aims by *engaging* their representations.' By 'representation' he means physical representation, and typically this will be discursive text, diagrams, or algebraic displays which are different 'representational types'. By 'engagement' Manders means generation and acceptance of these physical representations (following normal standards for them, of course). Contrary to the received wisdom which holds that diagrams are a potential source of error, Manders upholds the inferential practices involved in each of these different representational types, including making inferences based on geometric diagrams.

A diagram, a text, and an equation can all be about the same thing, yet can be decomposed in strikingly different ways. Different representations can bring out different aspects. These differences involve 'representational granularity', as Manders calls it. For example, a diagram showing the perpendicular to a base of a triangle must show that perpendicular either inside or outside the base. An equation describing that triangle with a perpendicular ignores whether it falls inside or outside. Thus, as Manders puts it, the diagram has a larger grain size than the equation. (However, this is only relative to the linkage between diagram and equation; without some suitable linkage representational types are simply non-comparable.) In traditional geometry, discursive text has smaller grain size than a diagram, and so will support much richer inferences. The exact length of a line, for instance, is not discernible from a diagram, but is easily captured in words. On the other hand, inferences based on, say, part/whole relations in a diagram are just as cogent as any piece of discursive reasoning. The 'control of representational grain size' is crucial, says Manders, to understanding many intellectual activities, but most especially to understanding mathematics.

Here's an interesting question: Does a particular notation *create* or merely *reveal* the properties of objects? Consider the example (mentioned by Manders) of a geometric figure, say a parabola, presented traditionally with a diagram or by means of an equation (see Figure 6.10). Does this curve have a degree? The idea would have made no sense at all to geometers before Descartes. But anyone trained in analytic geometry would say, Yes, it's a second-degree curve (because the equation describing a parabola, $y = ax^2$, has the number 2 for an exponent).

For the formalist-minded, who think that mathematical properties are essentially properties of notation, the property of 'degree' can only be strictly associated with the equation, not the diagram. Formalists may be happy to say the diagram of a parabola has degree 2, but that is because of the prior conventional association of the equation with the diagram. However, if analytic geometry had never been created, then the diagram would simply not have any degree at all.

A Platonist, on the other hand, instinctively says, Yes, the diagram itself does have a degree. The degree of a parabola only became *manifest* at the time of the creation of analytic geometry with its depiction of a parabola by means of a second-degree equation, but the relevant property was there all along.

It would seem that the Platonist has the upper hand, since formalists are in the embarrassing position of having to explain what it is about the diagram that made it possible to invent the notation of analytic geometry in such a way that this particular diagram turns out to have degree 2. In other words, formalists are going to have to postulate some objective property of the curve to explain why it is that analytic geometry works the way it does. In short, *notations reveal properties, they do not create them.*

This was an easy victory for Platonism over formalism. A much more interesting challenge comes from this question: Are there properties that can only be discovered using a particular notation? We might well grant that a notation does not create anything, it merely reveals what is already present, but we still wonder whether a specific notation was essential for making a particular discovery.

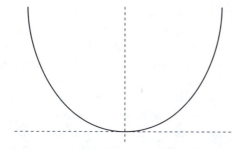

Figure 6.10 A parabola

The degree of a curve was not created by analytic geometry, but could the property of degree have been discovered without Descartes' form of representation? I suspect that the answer is No; particular types of notation are indeed essential for the discovery of certain properties. Microscopes did not create microbes, but may have been necessary for revealing them.

There is a theoretical reason for thinking that different notations would capture different truths. For many mathematical objects of interest there will be uncountably many relevant facts. Any reasonable notation for describing these facts would consist of a countable number of basic elements (e.g. an alphabet) and only allow countable combinations of these in constructing statements to represent the facts. So there could only be a countable infinity of these representations. Consequently, many facts would have to go unrepresented in that notation. This doesn't mean that some facts are unrepresentable in principle. A particular fact need only go unrepresented in a particular notation – it might well be captured by another. Indeed, for any fact, there is some possible notation in which it is represented. This sketchy argument suggests that different notations are indeed necessary to reveal different mathematical facts. But the argument is no more than a sketch. It would be much better to stick to real working examples; so let's return to knot theory.

What are the different forms of representation used in knot theory?

(1) *Pieces of string, rope, etc. joined at the end.* One might think: These are not *representations* of knots; these *are* knots. Not so. The knots of knot theory don't, for instance, have any definite length; they can have arbitrarily many knottings with arbitrarily many twists in an arbitrarily short strand. These are characteristics that no physical piece of string can have. Of course, actual pieces of rope are excellent representations of knots and, clearly, working with ropes and cords and threads suggested knot theory in the first place. Also, many important results have been attained by manipulating pieces of string. But the knots of knot theory are abstract entities. Similarly, numbers are not bananas, though fruit fondling may be a good way to learn elementary arithmetic.

(2) *Projections.* The diagrams that we typically use are projections of knots onto the plane. We've only looked at projections onto the Euclidean plane; others are possible. Much interesting work stems from investigating projections onto the surface of a sphere or a torus, and so on. Projections make crossings manifest, and this is the key to most significant results in knot theory. Projections allow us to define crossings, crossing numbers and, most importantly, they allow us to define Reidemeister moves.

(3) *The Dowker and Conway notations.* I'll lump these together, since they are both connected to knot projections. In each case the crossings are labelled, then the labels are manipulated to achieve striking results. At this point one might think that these notations are not really forms of representation, but are rather applications of number theory to knots. In particular, Conway's notation for knots (with its remarkable use of continued fractions) is an application of

already known properties of numbers to knots. I don't think this distinction can be as altogether clear as it would seem to be. As I mentioned above, in the case of the Arabic positional notation for the natural numbers, brilliant notations do not come *ex cathedra*. Developing a notation and learning about the objects named goes hand in hand. In the case of the natural numbers, their recursive properties are mirrored in the notation. And it is certain that no one first invented the Arabic numerals and then later discovered that the natural numbers form a recursive set. I suspect that the right answer in the case of knot theory is *both* – the natural numbers are both a notation, a labelling device, and they are being applied.

(4) *Polynomials.* These are also defined on knot projections. It is not known whether polynomials of knots can be defined directly on knots or only on the planar projections (see Lickorish and Millett 1988). If the later turns out to be the case, then, interestingly, this means that *one form of representation is dependent upon another*.

Above, I mentioned the curious way calculations of knot polynomials are carried out. We wrote down things that look like equations – indeed they are equations. But note also that pictures occur in these equations. For instance, $\langle \times \rangle = A \langle \rangle \rangle \langle \rangle + A^{-1} \langle \asymp \rangle$. This notation – so commonly used because it is so efficient in calculating polynomials – is certainly not like other verbal symbolic notations. It seems more like hieroglyphics or picture writing. We'd be hard pressed, for instance, sharply to separate syntax from semantics, since the knot projection that the equation represents appears right in the equation itself.

There are a few morals that can be drawn from these simple observations. Notice that an object might be described in, say, Cartesian or spherical coordinates. There is an easy way of transforming the description from one to the other. Solving an equation might be easier in one than the other, but they are representationally equivalent. The same cannot be said for the different forms of representation in knot theory. Each picks out different things. Of course, they overlap to some extent, but there are properties that one notation can describe that others cannot.

One of the most important morals to be drawn from knot theory is perhaps to admit a very deep kind of incompleteness in any form of representation. Mathematics is so rich that no form of representation can hope to capture all of it. Curves actually have the property of degree, which we stumbled upon only by inventing the algebraic notation of analytic geometry. Knots – though you'd never know it from playing with bits of string – have properties associated with continued fractions and polynomials. And we discover these properties only by inventing new notations which make them manifest. Spinoza thought that God/nature has infinitely many different properties, but that we are only aware of two, thought and extension. The moral to be drawn from knot theory is that knots (and all other mathematical entities) are like this: they, too, have indefinitely many different kinds of attributes, and sometimes we only uncover them as we find new ways of representing them.

There's one final moral to draw about how language works.

Sense, Reference and Something Else

Frege famously declared that meaningful expressions have a sense and a reference (1892). 'The morning star' and 'the evening star' both have the same *reference* (Venus), but they differ greatly in *sense*, their 'mode of presentation', as Frege put it. We need not worry about the problems of this theory, nor about its serious rivals. I want only to note that it misses something important, an ingredient that ought to be present in *any* theory of mathematical language.

In addition to the sense and the reference of a term, there is something else which I'll call 'computational role'. The name '2' has a sense (i.e. the natural number which is the successor of one), and it has a reference (i.e. the number two). But it also plays a computational role in the Arabic notation. The role it plays is built right into the recursive notation itself. It is very easy to miss this aspect of a meaningful expression, since most philosophy of language is built on examples like 'the morning star', 'the present king of France', 'water' and 'gold'. We don't calculate with any of these terms; they play no role in systematic computation. But when we reflect on the amazing power of the Arabic numerals and spend some time playing with the various ingenious notations invented for knot theory, something like computational power becomes evident.

It is often suggested that Frege's sense is a method or procedure for determining the reference. Dummett (1978a) holds such a view, for example, and so does Moschovakis (1990) who has developed this view at great length. Fregean sense, according to him, should be seen as an algorithm (and reference as value, output). Some connection should be apparent among Dummett's procedure, Moschovakis's algorithm, and what I'm calling computational role. But they cannot be the same thing. To see this, note that the sense of the term '2' and the sense of the term 'two' are the same; anyone who didn't know '2 = two' would simply be ignorant of English. As Frege noted, someone might not know 'the morning star = the evening star'. This latter truth reflects a great astronomical discovery; the terms have the same reference, but quite different senses, unlike '2' and 'two'. By contrast, 'the even prime number' has a different sense from '2' and 'two', though all three terms have the same reference. Any procedure or algorithm (connected to sense *à la* Dummett or Moschovakis) involving '2' is the same as that involving 'two'. However, we obviously cannot calculate with 'two' in the same way as we calculate with '2' or even with the Roman 'II'. The first of these has little or no computational power; 'II' is quite a bit better, and '2' is truly a marvel.

The notions of sense and reference are simply not enough to do full justice to the computational brilliance of a clever notation. Computational role must be included along with them as an objective feature of language, especially mathematical language. Of course, I've done little more than point out the importance of this feature of notation; a full development awaits.

The point of this chapter has been to make clear some of the features of mathematical notation and to remind ourselves of how important and how brilliant a good notation can be. There is nothing new in this. Mathematicians have always appreciated clever notations; but symbolism is usually seen as a tool – it's what the tool does that we really care about. Fair enough. But if we want a richer appreciation of mathematics, we should focus some of our energy on this remarkable tool – notation. Besides mathematics (and may be Chemistry), poetry alone works wonders with it.

Further Reading

There is virtually nothing written on this topic. Cajori's *A History of Mathematical Notations* might be useful, but it is best to simply look at mathematics at work. Chemistry also makes good use of its notation, so a look at any chemistry text might be helpful, as well. For good introductions to knot theory, see Adams, *The Knot Book* or Livingston, *Knot Theory*.

CHAPTER 7
What is a Definition?

I t's hard to imagine a subject more likely than *definition*s to bring yawns to the reader. Yet, it's a topic packed full of interesting and important issues, many of them central to how we understand mathematics. The nature of definition is not much discussed today for the simple reason that there is an official view which is completely dominant and apparently unproblematic.

The Official View

This can be found, for example, in *Principia Mathematica*:

> A definition is a declaration that a certain newly-introduced symbol or combination of symbols is to mean the same as a certain other combination of symbols of which the meaning is already known.
>
> It is to be observed that a definition is, strictly speaking, no part of the subject in which it occurs. For a definition is concerned wholly with symbols, not with what they symbolize. Moreover, it is not true or false, being an expression of a volition, not a proposition.
>
> (Whitehead and Russell 1927: 11)

The same view is sometimes expressed by saying that a definition must satisfy the two criteria of *eliminability* and *non-creativity*. We begin with undefined terms, called primitives; then we must always be able to replace any defined term in favour of primitive ones (eliminability) and no new theorems should be proven with the help of definitions that could not be proven without them (non-creativity).[1]

Set theory, for instance, has two primitives, *set* and *is a member of.* All other concepts are defined using these. (Logical terms are taken as already understood.) *Subset*, for instance, is defined this way:

$$A \subseteq B \leftrightarrow_{def} (\forall x)(x \in A \to x \in B)$$

Instead of using the short and simple subset notation (on the left), we could always resort to the longer form in primitives (on the right). There is a theorem of set theory that says the empty set is a subset of every set: $(\forall S)\phi \subseteq S$. We could express this theorem using primitives as: $(\forall S)(\forall x)(x \in \phi \to x \in S)$. Thus, the symbol '$\subseteq$' can be eliminated in favour of primitives with no loss of ability in expressing the content of any theorem. Its utility as an abbreviation is evident, but the content of the theorem can be expressed without it.

Mathematicians often introduce concepts in a sloppy way – sloppy, that is, according to the official view. Typical is the introduction of the empty set through a definition: $\phi =_{def} \{x: x \neq x\}$. A textbook author who does this then goes on to prove the theorem that the empty set is a subset of every set. The reason this is (rightly) considered sloppy is that it is a creative definition. In such a presentation, the theorem could not be proved without the definition. The proper way to do things on the official view is to introduce the empty set by means of a definition *and* to assert its existence separately in an axiom. These are logically quite distinct. The definition of the empty set by itself does not guarantee its existence any more than a definition of a unicorn guarantees that there is one.

There is much to be said for the official view. It imposes a great deal of clarity and order on mathematics. But it was hard won, and has only become orthodoxy in this century.

The Frege–Hilbert Debate

The official view of definitions with its insistence on *eliminability* and *non-creativity* did not fall out of the sky. It largely resulted from a debate almost a century ago when the giants fought over these issues. The debate followed on Hilbert's publication of *Foundations of Geometry* (1899). Hilbert remarked that key terms were being defined *contextually* by the axioms; Frege objected to this. In the exchange the official view emerged.

Hilbert opened his *Foundations of Geometry* with a 'definition' as he calls it:

> Consider three distinct sets of objects. Let the objects of the first set be called *points* and be denoted by A, B, C, . . .; let the objects of the second set be called *lines* and be denoted by a, b, c, . . .; let the objects of the third set be called *planes* and be denoted by α, β, γ, . . .
>
> (Hilbert 1899: 3)

This seems like no definition at all, but rather a picking out of primitives, *point*, *line* and *plane*, and fixing a notation for each.

Hilbert presents his axioms in distinct groups; there are eight Axioms of Incidence, four Axioms of Order, five of Congruence, one of Parallels, and two of Continuity. To convey a feel for Hilbert's mode of presentation, I'll reproduce part of the section on Axioms of Order. Note especially the beginning remarks on definition.

§3. Axiom Group II: Axioms of Order

The axioms of this group define the concept 'between' and by means of this concept the *ordering* of points on a line, in a plane, and in space is made possible.

DEFINITION. The points of a line stand in a certain relation to each other and for its description the word '*between*' will be specifically used.

II, 1. *If a point B lies between a point A and a point C then the points A, B, C are three distinct points of a line, and B then also lies between C and A.*

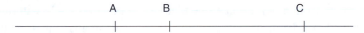

II, 2. *For two distinct points A and C, there always exists at least one point B on the line AC such that C lies between A and B.*

II, 3. *Of any three points on a line there exists no more than one that lies between the other two.*

(Hilbert 1899: 4f.)

There followed a series of letters between Hilbert and Frege, then two articles by Frege on the foundations of geometry (Frege 1971); he returned to the issue again a few years later in unpublished papers (Frege 1979). Many topics were discussed, but the nature of definitions and closely related matters was the principle focus of their exchange. Frege took issue with several claims and assumptions made by Hilbert. Among the specific topics are: definition versus explanation, contextual definition and the role of axioms, consistency and existence, independence proofs, and defining new versus old terms. Let's look at each of these.

Definition vs. Explication

We start with primitives or undefined terms. 'Every definition', says Frege, 'contains a sign (expression, word) which previously had no reference and which is given a reference only through this definition' (1971: 7). Axioms are expressed with primitives or defined terms. But axioms and definitions are not the whole of it. 'We may assume that there are propositions of yet a third kind: the explicatory propositions, which however, I should not like to consider as belonging to mathematics itself but instead should like to relegate to the pre-amble, to the propaedeutic' (*ibid.*: 8).

In any presentation of, say, set theory, 'set' and 'membership' are the primitives. Though undefined, they are much described. A textbook on set theory will typically say that a flock of birds or a pack of wolves is a set. If it's wise, it will further add that neither a flock of birds nor a pack of wolves is strictly a set, since a flock can fly south and a pack can be on the prowl, whereas sets go nowhere and menace no one (except the odd undergraduate). By means of such illustrative examples and informal discussion we *explicate* the concept of a set. Frege's point is that such an explication is part of the preamble, not part of mathematics proper.

After his exchange with Hilbert, Frege further clarified his view, giving a simple formulation of what is now the standard account: 'all [a definition] does in fact is to effect an alteration of expression. . . . it is not possible to prove something new from a definition alone that would be unprovable without it. . . . In fact considered from a logical point of view it stands out as something wholly inessential and dispensable' (1979: 208). As Frege notes, though, being without logical significance does not imply being without psychological significance. Definitions are very likely essential from a practical, human point of view.

Contextual Definition

Hilbert championed so-called contextual definitions. Terms are not explicitly and independently defined, but rather pick up their meaning by figuring in the axioms. Frege was dead set against this.

> It is absolutely essential for the rigour of mathematical investigations that the difference between definitions and all other propositions be maintained throughout in all its sharpness. The other propositions (axioms, principles, theorems) must contain no word (sign) whose sense and reference or (in the case of form-words, letters in formulae) whose contribution to the expression of the thought is not already completely settled, so there is no doubt about the sense of the proposition – about the thought expressed in it.
>
> (Frege 1971: 8)

The problem as Frege saw it was that if a term did not already have a sense before the statement of the axiom, then the axiom couldn't express a thought.

This didn't trouble Hilbert in the least. On the contrary, he saw it as a virtue. A formal theory applies to anything which satisfies the framework, not just the intended interpretation.

> [E]very theory is merely a framework or schema of concepts together with their necessary relations to one another, and that the basic elements can be construed as one pleases. If I think of my points as some system or other of things, e.g. the system of love, of law, or of chimney sweeps ... and then conceive of all my axioms as relations between these things, then my theorems, e.g. the Pythagorean one, will hold of these things as well.
>
> (Hilbert in Frege 1971: 13)

Hilbert's notion of contextual definitions is intimately tied to a more general theory of meaning which has been highly influential and which led to the infamous incommensurability of Kuhn, Feyerabend, and others. The doctrine of incommensurability says that as we change our beliefs (i.e. change our theory, axioms) we change the very meaning of the terms involved. Meaning and belief are intimately interconnected. As Hilbert puts it:

> [E]ach axiom contributes something to the definition, and therefore each new axiom alters the concept. 'Point' is always something different in Euclidean, non-Euclidean, Archimedean, and non-Archimedean geometry respectively.
>
> (Hilbert in Frege 1971: 13)

Interestingly, the same outlook has been proposed by prominent physicists in *Gravitation*, a very influential textbook on general relativity.

> [T]hat view is out of date which used to say 'define your terms before you proceed'. All the laws and theories of physics ... have this deep and subtle character, that they both define the concepts they use ... and make statements about these concepts. Contrariwise, the absence of some body of theory, law, and principle deprives one of the means properly to define or even use concepts. Any forward step in human knowledge is truly creative in this sense: that theory, concept, law, and method of measurement – forever inseparable – are born into the world in union.
>
> (Misner *et al.* 1973: 71)

Contextual definition and incommensurability go hand-in-hand.

Defining Old Terms

In setting up a formal system we often use terms which already have a well-established meaning. How should a definition of 'number' or 'addition', for instance, be understood? Since they have a sense already, it would appear that the notion of *arbitrary stipulation* is out of place. After his debate with Hilbert, Frege expressly worried about this problem (1979: 210–11). If we can give a correct analysis (using the primitives of the theory), so that both the old term and the new analysing expression have exactly the same sense, then the problem is solved. But we can never, of course, be sure that the two senses are the same. On the other hand, we might introduce a new term, say, 'knumber' (sounds like 'number'), which is defined by stipulation. And if this new term turns out to be completely adequate for all the purposes we could want the concept for, then we may simply forget about the old term 'number', since it is completely unnecessary. (This is often called eliminative reduction.) Again, problem solved. Either way, the difficulty presented by old terms is overcome; all definitions can indeed follow the pattern that we have been calling the official view.

Frege adds a crucial remark about analysing an old concept: 'The effect of the logical analysis of which we spoke will then be precisely this – to articulate the sense clearly. Work of this sort is very useful; it does not, however, form part of the construction of the system, but must take place beforehand' (1979: 211). In other words, the work that goes into this analysis is part of the preamble, the explication of the term; it is not part of mathematics proper.

Frege's account would seem to be completely unworkable in the light of Lakatos's example (which we shall look at below). The main point coming from Lakatos is that concepts are 'proof-generated', as he puts it. There is no old, pre-analytic concept of a polyhedron that is correctly analysed and defined in primitive terms, nor is there a new concept of polyhedron that can be so defined once and for all at the outset. Lakatos's point is that the concept of polyhedron changes as we theorize about it. On Frege's view, it must be fixed at the start. But this would seem to be impossible if Lakatos is remotely right.

Some of the greatest mathematical achievements would be judged 'pre-mathematical', by Frege's lights. Dedekind's theory of real numbers, for instance, would have to count as *not* mathematics proper. But then, what sort of activity is it? It is certainly not physics, nor is it literary criticism. It might conceivably be called philosophy, since it is concerned with foundational issues. But it seems absurd not also to call it mathematics.

The official view could only be maintained as some sort of regulative ideal – at the *end* of all mathematical theorizing. Then we can formulate the final system with definitions fixed at the outset. But this has nothing to do with *actual* mathematical activity, and any serious philosophy of mathematics must account, at least in broad outline, for how things are actually done.

Consistency and Existence

One of Hilbert's main aims was to show the consistency of his presentation of geometry. Frege failed to see the need of separate consistency proofs: 'From the fact that axioms are true, it follows that they do not contradict one another' (1971: 9). Hilbert couldn't be more opposed: rather than truth implying consistency, it's the other way round: 'If the arbitrarily posited axioms together with all their consequences do not contradict one another, then they are true and the things defined by these axioms exist. For me, this is the criterion of truth and existence' (Hilbert in Frege 1971: 12).

This is a very curious view, and Frege raised the obvious objection (1971: 18ff.): The notion of an all-powerful, all-loving, all-knowing being is (we may presume) a consistent concept. By Hilbert's lights we then have a proof of the existence of God, which seems preposterous.

Hilbert was not alone in holding this amazing view. Poincaré, too, held that consistency and existence go together. No sensible person thinks that the consistency of ghosts, goblins and the Loch Ness Monster (with everything else we know to be true) implies that these things exist. Yet, the analogous view in mathematics is maintained by some of the greatest mathematicians.

Poincaré's espousal of this principle seems something of a mystery. His philosophical sympathies run along constructivist or intuitionist lines, so one would expect him to link mathematical existence with constructability, not mere consistency. Hilbert is less surprising, if we think of him as a formalist. Then, mathematical existence is not taken in any serious literal sense. For example, suppose we are working with the real numbers and the statement 'every equation has a root' is consistent with everything else we maintain. Then it is true, and the roots exist, according to Hilbert. Thus, the equation $x^2 + 1 = 0$ has a solution, $x = \sqrt{-1}$. This entity, $\sqrt{-1}$, though previously unencountered and not itself a real number, is then taken to exist. Years later, when he became a formalist in full flight, Hilbert declared that we can add 'ideal elements' (points at infinity, imaginary numbers, transfinite numbers) to finite mathematics to make a smoother system. Hilbert's notion of existence is thus seen to be innocuous; it's a kind of fictional existence. It is certainly not the same sense we normally employ.

For Frege the Platonist, asserting mathematical existence is just as serious as asserting the existence of anything in the physical realm. And given the analogy that Platonists think holds between the mathematical and the physical, it is not surprising that he should dismiss Hilbert's linking of truth and existence with consistency.

But, perhaps Frege is too wedded to the analogy. At this point, the similarity between the physical and the abstract seems to me to break down. In the mathematical realm, it is plausible to think that possibility and actuality go hand-in-hand. Suppose all of *known* mathematics is captured by set theory. Now imagine that we posit some new sort of abstract entity, *zonks*, and we develop some intuitions about them, formulate axioms, prove some interesting theorems (including some with new consequences for set theory that we already know to be true), pose some interesting open problems about zonks, and so on. Assume, further, that all of this is shown to be independent of the axioms of set theory. It would be hard to deny that this, too, is legitimate mathematics. The axioms of zonk theory are just as likely to be true as set theory and zonks are just as likely to exist as sets. In short, if there is logical room for zonks, then they do exist. In the physical realm, *Occam's razor* is the battle cry; but in the Platonic realm: *the more the merrier*.

If this principle is right, then the official view's prohibition against so-called creative definitions is undermined. The interesting way in which it is violated comes (as in the graph theory examples which we will examine shortly) from alternative forms of representation. We introduce new concepts (face, unlabelled graph) and immediately start proving new results based on these concepts. It may be that these terms can be defined by the set theory primitives that graph theory adopts, but not in any feasible sense. They are, however, intuitively or antecedently understood – usually in terms of the diagram. We seem to have non-eliminable, creative definitions, and they play a major role in achieving some of graph theory's most important results.

Independence Proofs

One of Hilbert's main concerns in *Foundations of Geometry* was with showing that the axioms are independent of one another. He did this by constructing models which, first, make all the axioms but one, true, then, another model making all but that one axiom false. The isolated axiom is thereby shown to be independent of the others. Frege pounced on this. If terms are being defined by the axioms, then taking an axiom to be true in one setting and false in another changes the very meaning of any terms involved. As Hilbert earlier put it, the very meaning of 'point' differs in Euclidean and non-Euclidean geometry. So the parallel axiom is not one and the same thing which is being asserted one time and denied another in any proof of independence; the very meaning has changed. Thus, Hilbert's technique of showing independence won't work.

Elsewhere Frege remarks that 'What we prove is not a sentence, but a thought' (1979: 206). Thoughts (i.e. propositions) are perfectly definite, unambiguous things; Hilbert's axioms, which only contextually define key terms, allow an ambiguity which shows they cannot really be thoughts, the true subject matter of mathematics. Hilbert's way of providing independence proofs (i.e. reinterpretation of the axioms) brings this out fully.

Reductionism

The standard view of definitions does not impose reductionism upon us, but it certainly makes reduction easy and natural. In number theory, for example, we have several important concepts: *number, successor, prime, composite, perfect, square, even, odd*, and so on. However, all of these can be reduced to two: *number* and *successor*. In set theory, as I mentioned above, *set* and *is a member of* are the undefined terms; all other concepts, such as: *cardinal, ordinal, function, union, intersection, finite, infinite*, are defined by those two primitives. The standard view of definitions requires that all mathematical theories be presented in this way. Indeed, it requires that every theory in any intellectually respectable field be so formulated. There is no justification for taking *hydrogen atom* as something over and above what can be defined using the prior concepts: *proton, electron, central force*, etc. Within any theory there should be as few primitives as possible. But what about the relations among different theories?

It has long been an article of faith that all of mathematics can be reduced to set theory. This reduction involves two ingredients: First, all the concepts of theory T (say number theory) can be defined in terms of set theory concepts. This means that terms like number and successor which are primitives in number theory become defined terms in set theory. (For example, as mentioned earlier, one of the proposed definitions of particular numbers goes like this: $0 = \phi, 1 = \{\phi\}, 2 = \{\phi, \{\phi\}\}, \ldots$) The second ingredient of the reduction says: all of the theorems of T are theorems of set theory. That is, when the theorems of number theory are expressed in set theory terms, they can be proven from set theory axioms.

Logicists (Frege, Russell) hoped to carry out a further reduction of set theory to logic. This has generally been thought to be a failure, though much recent activity suggests that a significant chunk of the programme might succeed.[2] Regardless of the fate of logicism, the apparent reduction to set theory is a remarkable achievement. The reduction, such as it is, is not a concrete result sitting on the library shelf, something we can point to and say: There it is! To a large extent, it is a matter of faith. Much is done in, for example, Russell and Whitehead's *Principia Mathematica*. But much remains undone. The faith that it all can be reduced is not irrational dogma; when pushed, mathematicians usually do come up with set theory definitions of some new object that they wish to investigate. But a measure of scepticism, as we shall see, is not wholly out of place.

There are three attitudes we could adopt towards this alleged reduction of mathematics to set theory. First, it's true; all of mathematics really does reduce to set theory. Second, all of mathematics can be *represented* in set theory, but we should not think that mathematics = set theory. Functions, for example, can be represented as ordered pairs, but perhaps that is not what functions *really* are, and numbers have the same structure as the sequence of sets: ϕ, $\{\phi\}$, $\{\phi,\{\phi\}\}$, . . ., but that need not be what numbers *really* are. The claim may be analogous to one we sometimes encounter in discussions of the mind–body problem: the mental and the physical are correlated, but they are not identical. Third, there is simply no reduction at all; either some of the concepts of T cannot be plausibly defined by set theory concepts, or (even if they can) some of the theorems of T cannot be derived from the axioms of set theory.

As we continue to discuss definitions, we shall keep in mind this larger issue of reductionism.

Graph Theory

Graphs are defined as follows: A *graph*, G, is an ordered pair $\langle V,E \rangle$ such that the set E (of 'edges') is a subset of the unordered pairs of the set V (of 'vertices'). It is often added that V and E are finite. For example, G = $\langle \{a,b,c\}, \{\{a,b\}, \{b,c\}, \{c,a\}\} \rangle$ is a graph with three vertices and three edges. G' = $\langle \{a,b,c,d,e\}, \{\{a,b\}, \{b,c\}, \{b,d\}\} \rangle$ has five vertices and three edges. Note that the basic definition of a graph is given completely in set theory terms.[3]

No sooner are such definitions and examples given, than the typical text on graph theory says that it would be natural to show a picture of the graphs. Our two examples, G and G', look like this:

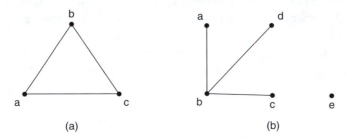

(a)   (b)

Figure 7.1 The graphs G (left) and G' (right)

Looking at both the set theory definition and at the diagram, it is little wonder that working mathematicians (as well as the rest of us onlookers) have a strong preference for the picture. Nevertheless, the set theory definition proves to be very useful, for example, in letting us know that Figure 7.2 is a picture of the same graph as Figure 7.1(b), in spite of their very different appearances.

A *planar graph* is one which can be drawn with no edges crossing. The graph (known as K_4) in Figure 7.3(a) can be redrawn as in Figure 7.3(b). (K_4 is called a *complete* graph because each vertex is joined to every other vertex.) The so-called utility graph (Figure 7.4) is not a planar graph. (Try redrawing it

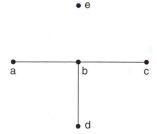

Figure 7.2

without any crossing lines; you will quickly become convinced that it cannot be done.)

There are different ways of proving a graph non-planar. One way uses the Jordan Curve Theorem which says that any simple closed curve divides the plane into two regions and that any continuous line from inside to outside cuts the boundary. The ideas of crossings, being planar, and the use of the Jordan Curve Theorem, etc. all make essential use of the diagrammatic representation of graphs.

A fundamental theorem in graph theory is Euler's theorem which we saw earlier in connection with polyhedra (and will see again below). In this setting it says: $V - E + F = 2$ for any planar, connected graph (where V, E and F are the number of vertices, edges and faces, respectively; the region outside the graph is considered a face; *connected* means no vertex is isolated as, for example, e is an isolated vertex in Figure 7.1(b)). Figure 7.5 illustrates the Euler relation.

'Vertex' and 'edge' were defined set theoretically. The notion of a 'face' can be, too – but not readily. It is obviously a geometric notion, and Euler's theorem is completely geometric in spirit.

Two graphs are *isomorphic* when there is a one–one correspondence between their vertex sets, a correspondence which preserves adjacency. Thus, in Figure. 7.6, the graphs G and G' are isomorphic.

(a)

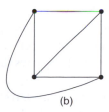

(b)

Figure 7.3

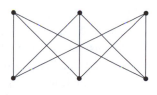

Figure 7.4

Figures 7.3 and 7.4 The graph 7.3(a) is planar since it can be redrawn as 7.3(b); 7.4 is not planar

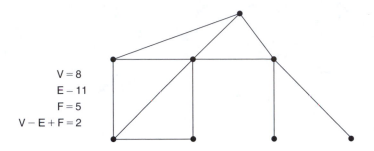

$$V = 8$$
$$E - 11$$
$$F = 5$$
$$V - E + F = 2$$

Figure 7.5 An illustration of the Euler relation

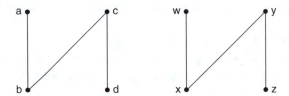

Figure 7.6 Isomorphic graphs

It is common to identify isomorphic graphs and simply say that $G = G'$. If we look at the diagrams, this identification is perfectly natural. Given the set-theoretic definition of graphs, however, we must say that strictly speaking G and G' are members of the same equivalence class (the isomorphism providing the equivalence). This seems a straightforward matter and quite unproblematic until we consider the notion of an *unlabelled graph*. Notice that in the figures above, some of the graphs had labels (names attached to the vertices) and some did not. Given the set-theoretic definition of a graph, what could an unlabelled graph possibly be? Graph theory books and research papers are full of unlabelled graphs, but they are nowhere defined.[4]

The most natural possibility that springs to mind runs something like this: An unlabelled graph diagram is a picture of an equivalence class of all labelled graphs which are isomorphic to some particular labelled graph. This definition begins to look implausible, however, when we consider the automorphisms of an unlabelled graph. (An *automorphism* is an isomorphism of the graph to itself.) Consider first, the automorphisms of the labelled graph in Figure 7.7(a). One is trivial, the identity map where everything stays the same. Another corresponds to a rotation of $\frac{2}{3}\pi$ radians. Yet another corresponds to a reflection where a and c change places. These are easy to characterize set-theoretically. The identity transformation is: $a \mapsto a, b \mapsto b, c \mapsto c$. In the rotation we have $a \mapsto b$, $b \mapsto c, c \mapsto a$, the set of edges is unchanged. In the reflection example, $a \mapsto c$, $c \mapsto a, b \mapsto b$; the set of edges is again unchanged.

Now consider the unlabelled graph pictured in Figure 7.7(b). The three automorphisms I mentioned in Figure 7.7(a) (identity, rotation, reflection) are easily

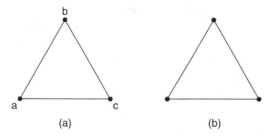

Figure 7.7 Labelled and unlabelled graphs

grasped here as well. These automorphisms are intuitively straightforward ideas when applied to the *picture* of the unlabelled graph. But how should they be characterized in terms of set theory? We might wonder: can't we just assume there are labels on the vertices, labels which are defined by the equivalence class of isomorphic graphs? This seems an obvious solution to our problem, so let's try it. Consider the set of graphs equivalent to G, and note that G has the vertex set {a,b,c}; let's call the equivalence class [[G]] and say that the unlabelled graph has the 'hidden' labels [[a]], [[b]], and [[c]]. Initially this looks attractive, but since any rotation is isomorphic to the initial graph, it follows that [[a]] = [[b]] = [[c]], so we lose our distinct vertices. The 'obvious' solution will not work.

I dare say there is an appropriate set-theoretic definition of automorphism for unlabelled graphs, but it will prove no easy matter coming up with something. I doubt that any definition which could be constructed in accord with the official view will be either natural or useful. It is no surprise that graph theory books do not provide one. Rather they all rely on a concept of unlabelled graph which is clearly taken from graph diagrams. The unlabelled graph – which, I stress, is a picturable geometric entity – would seem to be the primary concept. A labelled graph is then understood to be an unlabelled graph with labels attached. This is exactly the reverse of what is implied by defining graphs set-theoretically. In the official way of doing things, we start from labelled graphs, then abstract from them to arrive at the unlabelled ones.

I hope this doesn't seem like a simple confusion of objects with their names. Such a confusion would be to fail to distinguish the object, a, from it's name, 'a'. In calling a graph unlabelled I mean to say more than a vertex has no name. I mean also to say there is no object there. Of course, for some graphs the set of vertices might be people, {Alice, Bob, and Carol}, or a set of events, {E, F, G}. By unla-belled graph I mean to abstract away not only the names 'Alice' or 'E', and so on, but also from the objects Alice and E, and so on. You might think of the graph the way structuralists think of mathematical objects in general – they are structures and the vertices are mere placeholders; there is nothing else there. Though I do not find structuralism in the philosophy of mathematics generally appealing, some-thing like it seems quite right in making sense of an unlabelled graph.

In thinking about this I've drawn heavily on conversations with Alasdair Urquhart who remarks: 'There is a mismatch between intuitive combinatorial mathematics and set theory. The set theory universe *imposes* a label on everything, so the labelled objects are primary by fiat. But this makes even elementary combinatorial mathematics clumsy. So it seems to me that such considerations throw some doubt on the "mathematics = set theory" equation.'

In principle there may be a set theoretic characterization of everything that goes on in graph theory – but providing such definitions and working with them is in no way feasible. For the most part graph theorists do not use set theory to make discoveries, nor to construct proofs, nor to make presentations of their results for all the world to see.

Having said this, I must now state a significant caveat, but one which reinforces the main point. When graph problems are set up for computers, the geometric notions must be defined in set theory terms. This would seem to support the underlying primacy of set theory. However – and this is the crucial point – mathematicians do not work with these definitions; they are only for machine computation. The *working* definition is based on the picture, the diagram.

On the other hand, the set-theoretic characterization is not a useless, artificial encumbrance – it is what makes different pictures representations of the same graph, and it plays a major role in proofs of some highly important theorems. The main results of Ramsey theory and random graphs, for example, are almost wholly in set-theoretic terms. Given a collection of any six people, at least three of them will be mutual acquaintances or complete strangers. That's a special case of Ramsey's theorem. The situation can be captured in graph theory terms by letting the people be vertices and the acquaintance relation and the stranger relation be distinguished edges. The proof of the theorem is carried out in set theory. A picture of the graph helps us to see what's at issue, but does not show that the theorem is true or give any hint as to its proof. The theory of random graphs, as you might imagine, makes heavy use of probability theory which, of course, is a branch of set theory. In browsing through graph theory textbooks, it's striking to notice the sudden drop-off of diagrams when random graphs are discussed. Bollobás (1979), for example, has not a single picture in his chapter on random graphs. Up to that point he was averaging almost one per page.

The problems that graph theory presents for the standard account of definitions stem from the fact that graph theory makes heavy use of two distinct representations of graphs – the set-theoretic and the pictorial. Basic concepts such as *vertex* and *edge* are readily captured in both representations. Others, such as *unlabelled graph* and *face* are natural in the pictorial representation, but forced in the set-theoretic; *Ramsey number*, on the other hand, is naturally set-theoretic and only artificially characterized pictorially.

Before drawing morals in any detail, let's consider a different source of difficulty for the official view of definitions, this time from Lakatos's account of the development of mathematics.

Lakatos

The standard account of definitions is totally foreign to the mathematical world of Imre Lakatos. In his wonderful *Proofs and Refutations* (1976) Lakatos, as I mentioned earlier, retraces the history of a theorem, the Descartes–Euler claim that for any polyhedron V − E + F = 2. As far as examples go, it's just a co-incidence that this theorem was mentioned above in my discussion of graph theory and again with polyhedra. But it is not really such a coincidence when you consider the type of examples involved – those with both algebraic and geometric features prominent.

Recall that *Proofs and Refutations* is a dialogue in which various characters stand in for historical figures. As the story unfolds, the theorem is proven, but counter-examples arise; definitions are proposed and revised, and so on. Let us enter Lakatos's dialogue in the middle, at a point where a counter-example has been given to the initial conjecture that V − E + F = 2 (which, at this point in the story, has already been 'proven'). The counter-example is a pair of nested cubes (Figure 7.8), in which V − E + F = 4.

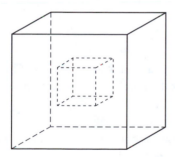

Figure 7.8

DELTA: But why accept the counter-example? We proved our conjecture – now it is a theorem. I admit that it clashes with this so-called 'counter-example'. One of them has to give way. But why should the theorem give way, when it has been proved? It is the 'criticism' that should retreat. It is fake criticism. This pair of nested cubes is not a counter-example at all. It is a *monster*, a pathological case, not a counter-example.

GAMMA: Why not? *A polyhedron is a solid whose surface consists of polygonal faces*. And my counter-example is a solid bounded by polygonal faces.

TEACHER: Let us call this *Def. 1*.

DELTA: Your definition is incorrect. A polyhedron must be a *surface*: it has faces, edges, vertices, it can be deformed, stretched out on a blackboard, and has nothing to do with the concept of 'solid'. *A polyhedron is a surface consisting of a system of polygons.*

TEACHER: Call this *Def. 2*.

DELTA: So really you showed us *two* polyhedra - *two* surfaces, one completely inside the other. A woman with a child in her womb is not a counter-example to the thesis that human beings have one head.

ALPHA: So! My counter-example has bred a new concept of polyhedron. Or do you dare to assert that by polyhedron you *always* meant surface?

TEACHER: For the moment let us accept Delta's *Def. 2*. Can you refute our conjecture now if by polyhedron we meant a surface?

ALPHA: Certainly. Take two tetrahedra which have an edge in common (Figure

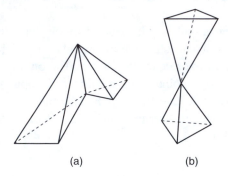

(a)   (b)

Figure 7.9

7.9(a)). Or, take two tetrahedra which have a vertex in common (Figure 7.9(b)). Both these twins are connected, both constitute one single surface. And, you may check that for both $V - E + F = 3$.

TEACHER: *Counter-examples 2a and 2b*.

DELTA: I admire your perverted imagination, but of course I did not mean that *any* system of polygons is a polyhedron. By polyhedron I meant *a system of polygons arranged in such a way that (1) exactly two polygons meet at every edge and (2) it is possible to get from the inside of any polygon to the inside of any other polygon by a route which never crosses any edge at a vertex*. Your first twins will be excluded by the first criterion in my definition, your second twins by the second criterion.

TEACHER: *Def. 3*.

(Lakatos 1976: 14–15)

Such activity is typical in the history of mathematics and Lakatos (with qualifications) endorses it. In this he may be extreme, but he is not entirely alone. Many proponents of the standard account of definition say that not only should a definition satisfy the above criteria of eliminability and non-creativity, but a definition should also (where appropriate) be an adequate explication of an intuitive or pre-analytic idea. The issue came up in the Hilbert–Frege debate. According to Whitehead and Russell's *Principia Mathematica*:

In spite of the fact that definitions are theoretically superfluous, it is nevertheless true that they often convey more important information than is contained in the propositions in which they are used. [For example, as] when what is defined is (as often occurs) something already familiar, such as cardinal or ordinal numbers, the definition contains an analysis of a common idea, and may therefore express a notable advance.

(Whitehead and Russell 1927: 11–12)

The debate above over the proper definition of polyhedron is a quarrel over the right explication of the pre-analytic or intuitive idea of a polyhedron. Champions of the standard view could well accept this sort of activity. They (with Frege) might think of it as laying the groundwork for the development and presentation of a formalized theory. However, the implicit view is that once the pre-analytic concept has been nailed down, then the theory may be cast in canonical form with all terms defined by means of the primitives; the conditions of *eliminability* and *non-creativity* would then be satisfied thereafter. Standing *outside* the system, we can pass judgement and say that a particular definition does or does not capture the intuitive, pre-analytic concept.

The account sounds plausible, but it does not begin to touch Lakatos's principle point. As an activity, analysing pre-analytic concepts in order to recast them in a precise or canonical form is essentially conservative. He has a much more radical account of concept formation. According to Lakatos, the best way to get better definitions is through proofs. He is something of an essentialist in that he adopts an Aristotelian view involving 'real' definitions, definitions which are not merely 'nominal' or stipulative, but are actually true or false. On the other hand, his definitions are not required to capture our pre-analytic, intuitive notions (which makes him quite un-Aristotelian). Lakatos's position is rather novel: *definitions are theoretical.*

One reason we keep modifying them is simply that the definitions we propose are fallible attempts to capture our intuitive concepts. That's the more conservative enterprise, the one that can be incorporated into the standard account of definitions. But, says Lakatos, there is a second, more important reason: theorizing actually *changes* our concepts. No concept is static; we shall always have to modify our existing definitions, as he sees it, since *conceptual change is an inevitable by-product of theorizing.*

PI: *Proof-generated concepts* are neither 'specifications' nor 'generalizations' of naive concepts. The impact of proofs and refutations on naive concepts is much more revolutionary than that: they *erase* the crucial naive concepts completely and replace them by proof-generated concepts. . . . In the different proof-generated theorems we have nothing of the naive concept. That disappeared without a trace. . . . The old problem disappeared, new ones emerged. After Columbus one should not be surprised if *one does not solve the problem one has set out to solve.*

(Lakatos 1976: 89–90)

Definitions are conjectures. They are declarative sentences, which – if correct – assert matters of fact. They are also subject to revision as a result of future theorizing, either because the initial formulation was wrong or because *the concept itself* has changed in the meantime. Since mathematics does not have a foundation, according to Lakatos, there are no primitive terms (terms properly singled out as undefined), and so defined expressions cannot be 'eliminated' (unpacked into a preferred set of primitives). Further, the definitions are obviously 'creative' since we can now derive things with the help of a definition which we could not derive otherwise. Neither of the standard criteria for definitions is satisfied in the Lakatosian way of doing mathematics, which is to say, the way things have often actually been done and the way Lakatos thinks they should be done.

The distinction between definitions and theorems is blurred. Logically, they are on a par – both have a truth value and both could be overthrown in the light of further evidence. The difference is methodological – only theorems are proved. Such is the lesson of history, according to Lakatos, and if we want better mathematics, we had better start letting proofs generate the definitions for us; we should abandon the strict insistence on nominal definitions.

Lakatos makes his case in a convincing way using the early history of polyhedra. Perhaps we should distinguish between young and mature theories, and say that the standard criteria of eliminability and non-creativity hold for *mature* theories. Lakatos, it must be admitted, has an ambivalent attitude towards this. At times he distinguishes between the two and qualifies the heuristic method of proofs and refutations as being applicable only to growing theories (1976: 42), to what he calls 'informal, quasi-empirical mathematics' (1976: 5). But at other times he points out that even mature theories can be rejuvenated. This (rightly) suggests that the distinction between growing and mature theories, if it exists at all, is blurred.

Not too much should be made of the fact that Lakatos's example is a highly intuitive, geometric one. Similar things can be said about several other branches of mathematics. Recall, for instance, what was said earlier about the notion of a set. Current axiomatizations of set theory have *set* and *membership* as primitives, but set theory, nevertheless, tries to capture the correct idea of a set. Cantor expressed his original conception as follows: 'By a "set" we understand every collection to a whole *M* of definite, well-differentiated objects *m* of our intuition or our thought' (Cantor 1895: 282). This conception is more or less destroyed by Russell's Paradox (since there is no set of objects corresponding to the thought 'is not a member of itself') and by the Axiom of Choice (since there are sets corresponding to no intuition or thought). The current conception of set is based on the so-called cumulative hierarchy. We start with the empty set, then build up the hierarchy with arbitrary applications of the axiom of unions, the power set axiom, etc. While current set theory is very impressive, most set theorists who work on foundations claim that the concept of set is really not well understood. And they think that only with a great deal of further research into so-called large cardinals and other exotica, will a better understanding come. The concept of set could well change again as a result of further research.

The moral from set theory is the same as the moral from polyhedra: There is no pre-analytic concept that we can first fix and then from that perfectly secure point proceed to do proper mathematics. Theorizing has changed the concept of set radically and could do so again. (Over the past 100 years the concept of *electron* has similarly evolved as a result of theorizing.) This is a process which happens *after* the pre-analytic concept has been incorporated either as an undefined primitive or as a defined notion in accordance with the criteria of the official account of definitions. There is no reason to think that there is ever a time when a concept is finally stabilized.

Concluding Remarks

The official view of definitions has much to recommend it. It clarified an enormous number of confusions and its imposition on working mathematics was a real advance. But it cannot be completely right. Some concepts (polyhedron) have a history and some theories (graph theory) have multiple representations (set-theoretic and pictorial). The official view cannot cope well with either of these. A quite different approach to mathematical definitions is needed. The question in the title – what is a definition? – remains. It is a wide-open problem.

Further Reading

An old, but useful book on the nature of definition in general is Robinson, *Definition*. The standard account of definition within mathematics is nicely presented in Chapter 8 of Suppes, *Introduction to Logic*. Bollobas, *Graph Theory* is a standard work on that topic. Lakatos, *Proofs and Refutations* has much to say of great interest on a number of topics, including definitions.

CHAPTER 8
Constructive Approaches

Let's define a number p as follows:

$$p = \begin{cases} 3, \text{ if Goldbach's conjecture is true} \\ 5, \text{ if Goldbach's conjecture is not true} \end{cases}$$

Now here's a simple question: Is p a prime number? The obvious and natural thing to say is Yes. And the proof is utterly trivial. The numbers 3 and 5 are both prime, and Goldbach's conjecture is either true or false, so either way p is a prime. QED.

Amazingly, constructivists would not accept this. We cannot assert something unless we have a proof – a *constructive* proof. If p is indeed a prime, then it is some particular prime. If it is 3, then we need a proof of this. But we could only produce a proof that $p = 3$ by producing a proof of Goldbach's conjecture, something that we cannot now do. Nor can we now refute Goldbach's conjecture, so we can't prove that $p = 5$, either.

You may find this absurd – I do. The natural response is to say that if constructive approaches lead to this sort of thing then let's chuck them out. But not everyone finds this absurd. As someone once said in a different context: 'That's not a problem – that's my theory.'

Let's focus on a quite different kind of example, one which will likely promote a measure of sympathy for constructivism.

In one of Shakespeare's greatest creations, *Hamlet*, we learn that the Prince of Denmark was a young man, that he was upset with his father's death, horrified with his mother's remarriage, in love with Ophelia, a friend of Horatio, good with a sword, and many other things. We can ask questions about Hamlet such as, Was he indecisive? The standard answer is Yes, and various facts mentioned in the play are cited in support. But what about questions concerning Hamlet's great-grandfather? Did he have blue eyes? Was he indecisive, too? Our instinctive answer is that these are meaningless questions. Shakespeare didn't mention such a character in the play, so there seems to be *no fact of the matter* about the eye colour or anything else having to do with Hamlet's great-grandfather.

There's a sharp contrast between the case of Hamlet's great-grandfather's eye colour and my great-grandfather's eye colour. The latter actually existed. I have a drawing of him, but it's in charcoal, so the eye colour is not revealed. No living member of the family knows the colour and it's not mentioned in any surviving letter, etc. Nevertheless, there would seem to be a fact of the matter about his eye colour, though I may never know what it is. But as for Hamlet's great-grandfather's eye colour, there simply isn't a fact to be known. An omniscient being would know the eye colour of my great-grandfather, but not of Hamlet's.

With this example in mind, let's redefine our number p as follows:

$$p = \begin{cases} 3, \text{ if Hamlet's great-grandfather has blue eyes} \\ 5, \text{ if Hamlet's great-grandfather does not have blue eyes} \end{cases}$$

In this case, it does not seem so preposterous to say that p is not defined. So resisting the question, Is p a prime?, isn't altogether nutty, after all.

There's a long tradition of thinking of mathematical objects as mental entities. They are created by the mind, just as Hamlet and other fictional characters are. If the mind has not got around to creating them, then, like the eye colour of Hamlet's great-grandfather, there is nothing to be known. The analogy is far from perfect, but it will greatly help in coming to understand many of the peculiarities of constructivism in mathematics.

From Kant to Brouwer

The grandfather of modern constructive mathematics is Immanuel Kant. One of Kant's central doctrines is that we do not experience things as they are independently of us, but rather that the mind provides much of the framework of experience. Space, time and causal relations, for instance, are supplied by us. They are not part of an independent, objective reality – something which Kant and his followers find utterly inconceivable. We experience objects as having a location in space and events as happening in time, but that's because space and time are the mind's contribution to experience – they are the *form* of experience. Kant's view of mathematics is based on this. Our a priori knowledge of geometric truths stems from the fact that space is our own creation. And arithmetic, according to Kant, is connected to our perception of time. The crucial element is the perception of succession. And so our a priori knowledge of numerical truths stems from the fact that time is our own mental creation.

L.E.J. Brouwer took Kant to be profoundly right about arithmetic (and more generally, to be right about algebra and analysis, which he believed to be based on arithmetic). Interestingly, he thought the development of non-Euclidean geometry showed that Kant's account of geometry was wrong.

Let me digress a moment. It's interesting to see the strong relation to Kant had by most of the main players in the philosophy of mathematics. Hilbert, the formalist, was a thoroughgoing Kantian about finite arithmetic, to which he added 'ideal elements' to recover all of classical mathematics; Frege, the logicist and Platonist, rejected Kant's account of arithmetic, but completely accepted Kant's view of geometry; Brouwer, the constructivist, reversed this, embracing Kant's view of arithmetic, but rejecting his account of geometry. Even Russell characterized some of his views with respect to Kant. There is one passage which never fails to shock me. In his early logicist days (when he first adopted the view that mathematics = logic), Russell said, 'Kant never doubted that the propositions of logic are analytic, whereas he rightly perceived that those of mathematics are synthetic. It has since appeared that logic is just as synthetic as all other kinds of truth' (Russell 1903: 457). One could, I suspect, reconstruct a great deal of the philosophy of mathematics simply in terms of attitudes and reactions to Kant.

If Kant was the grandfather, then Brouwer is the father of modern constructive mathematics. Current constructivism has a greater debt to his intuitionism than to anything else, so let's now have a look at it. Needless to say, since there are many construtivists, there are many versions of constructivism. I will try to give the general idea, ignoring subtle differences.

Brouwer's Intuitionism

In fine Kantian fashion, Brouwer

> considers the falling apart of moments of life into qualitatively different parts, to be reunited only while remaining separated by time, as the fundamental phenomenon of the human intellect, passing by abstracting from its emotional content into the fundamental phenomenon of mathematical thinking, the intuition of the bare two-oneness.
>
> (Brouwer 1913: 80)

Continuing in the same vein, Brouwer writes:

> This intuition of two-oneness, the basal intuition of mathematics, creates not only the numbers one and two, but also all finite ordinal numbers, inasmuch as one of the elements of the two-oneness may be

thought of as a new two-oneness, which process may be repeated indefinitely; this gives rise still further to the smallest infinite ordinal number ω. Finally this basal intuition of mathematics, in which the connected and the separate, the continuous and the discrete are united, gives rise immediately to the intuition of the linear continuum, i.e. of the 'between', which is not exhaustible by the interposition of new units and which therefore can never be thought of as a mere collection of units.

(*ibid.*: 80)

This is pretty obscure stuff. It was written early in his career. Alas, four decades later it got no better. The 'first act of intuitionism', says Brouwer

completely separates mathematics from mathematical language, in particular from the phenomena of language which are described by theoretical logic, and recognizes that intuitionist mathematics is an essentially languageless activity of the mind having its origin in the perception of a move of time, i.e. of the falling apart of a life moment into two distinct things, one of which gives way to the other, but is retained in memory. If the two-ity thus born is divested of all quality, there remains the empty form of the common substratum of all two-ities. It is this common substratum, this empty form, which is the basic intuition of mathematics.

(Brouwer 1952: 141–42)

There is also a 'second act of intuitionism' according to Brouwer, 'which recognizes the possibility of generating new mathematical entities',

firstly, in the form of infinitely proceeding sequences $p_1, p_2, \ldots$, whose terms are chosen more or less freely from mathematical entities previously acquired . . .;

secondly in the form of mathematical species, i.e. properties supposable for mathematical entities previously acquired, and satisfying the condition that, if they hold for a certain mathematical entity, they also hold for all mathematical entities which have been defined to be equal to it.

(*ibid.*: 142)

Mathematics, for Brouwer, is first and foremost an *activity*. Mathematicians do not discover pre-existing things, as the Platonist holds and they do not manipulate symbols, as the formalist holds. Instead, mathematicians make things. It is a languageless activity, just as is doing physics or eating lunch.

Brouwer stresses this, since formalism is essentially about symbols, language. Perhaps the best analogy is with baking a cake. If I'm successful, I can describe my non-linguistic activities of mixing the ingredients, setting the oven temperature, etc. in the form of a recipe, which, of course, is a linguistic entity. But the recipe is not the activity of baking a cake; it is merely an aide to others who might want to bake a similar cake for themselves.

Bishop's Constructivism

I'll do no more than cite a few key passages from Errett Bishop whose main contribution is in the mathematical details. His *Foundations of Constructive Analysis*, originally published in 1967, is a technical *tour de force*. After his death Douglas Bridges brought out a revised and re-named edition, *Constructive Analysis* (1985), with the same 'Manifesto' from which I'll quote.

> The primary concern of mathematics is number, and this means the positive integers. . . . Mathematics belongs to man, not God. We are not interested in properties of the positive integers that have no descriptive meaning for finite man. When a man proves a positive integer exists, he should show how to find it. If God has mathematics of his own that needs to be done, let him do it himself.
>
> (Bishop and Bridges 1985, 4–5)

> Building on the positive integers, weaving a web of ever more sets and more functions, we get the basis structures of mathematics . . . Everything attaches itself to number, and every mathematical statement ultimately expresses the fact that if we perform certain computations within the set of integers, we shall get certain results. . . . even the most abstract mathematical statement has a computational basis.
>
> (*ibid.*: 6–7)

> The transcendence of mathematics demands that it should not be confined to computations that I can perform, or you can perform, or 100 men working 100 years with 100 digital computers can perform. Any computation that can be performed by a finite intelligence – any computation that has a finite number of steps – is permissible.
>
> (*ibid.*: 6)

In sum, to be meaningful, according to Bishop, mathematics must be accessible to humans, and this in turn means computable in a finite number of steps with a result expressing a relation among numbers. On the side of liberality, Bishop allows that this need only be *in principle*; questions of what is practically computable are of no concern.

Dummett's Anti-realism

Brouwer and Bishop have been the mathematicians most prominent in advocating constructivism. Michael Dummett is the philosopher who has done most to promote this approach. He is also, unquestionably, the leading Frege scholar and one of the most influential philosophers of mathematics today.

Dummett's views on logic and language, metaphysics and mathematics form a seamless whole. *Semantic anti-realism* is Dummett's doctrine that we need not take every well-formed statement as having a determinate truth-value. *Realism*, by contrast, asserts that every statement is true or false, and this truth or falsity is independent of us and how we might go about checking this truth-value. Dummett's views are quite general; they stem from considerations in several diverse fields, including statements about the past, counter-factuals, and, of course, mathematics.

Dummett asks us to imagine a man, Jones, now dead, who in life *never* encountered any danger. Consider the sentence: 'Jones was brave.' Does it have a truth-value (of which we may be forever ignorant)? If Yes, then there must be facts about Jones which would make the following counter-factual sentence true (or make it false): 'If Jones was in a dangerous situation, he would have acted bravely.' But suppose there are no facts about his character, his brain structure, and so on that we could single out which determine the truth of the counter-factual. Dummett remarks:

> [I]f such a statement as 'Jones was brave' is true, it must be true in virtue of the sort of fact we have been taught to regard as justifying us in asserting it. It cannot be true in virtue of a fact of some quite different sort of which we have no direct knowledge, for otherwise the statement 'Jones was brave' would not have the meaning that *we* have given it.'
>
> (Dummett 1959a: 16)

In accepting this, says Dummett, we make 'a small retreat from realism; [we] abandon a realist view of character' (*ibid.*). In doing so, we abandon bivalence; the statement 'Jones was brave or Jones was not brave' cannot be asserted come what may.

Dummett talks about 'the meaning *we* have given it'. This is a Wittgensteinian idea – meaning and use are linked. How we learn certain concepts is central, and it has deep implications for how we do (or at least ought to do), mathematics. The commonly accepted account is that the meaning of some particular sentence, *S*, is grasped by knowing its truth-conditions, that is, by knowing how things would be in order for *S* to be true. Thus, we understand 'Raven 1 is black' when we know how the world must be in order for this sentence to be true. Similarly, to understand the meaning of 'Raven 2 is red' or 'Raven 3 is studying for a logic test' we would have to know how things must be in order for these to be true. Compound sentences are built out of previously understood components: thus, 'Raven 1 is black and raven 2 is red.' Similarly, quantified sentences should be no problem: thus, 'All ravens are black.' As long as we comprehend the components or instances in terms of their particular truth-conditions, we understand the more complex statements. However, Dummett emphatically opposes this commonplace view. He says we learn about, say, the universal quantifier by learning how to *use* it. 'We learn to assert "For all *n*, F*n*" when we can assert "F(0)" and "F(1)" and . . .' Consequently, Dummett holds, meaning is not given by truth conditions. '*We no longer explain the sense of a statement by stipulating its truth-value in terms of the truth-values of its constituents, but by stipulating when it may be asserted in terms of the conditions under which its constituents may be asserted*' (1959a: 18, his italics).

What goes for Jones's character goes for mathematics.

> The sense of, e.g., the existential quantifier is determined by considering what sort of fact makes an existential statement true, and this means: the sort of fact which we have been taught to regard as justifying us in asserting an existential statement. What would make the statement that there exists an odd perfect number true would be some particular number's being both odd and perfect;
>
> (Dummett 1959a: 17)

This seems quite right. But Dummett then goes on to state what certainly looks like a non sequitur:

> hence the assertion of the existential statement must be taken as a claim to be able to assert some one of the singular statements. We are thus justified in asserting that there is a number with a certain property only if we have a method for finding a particular number with that property.
>
> (*ibid.*)

He continues further and draws the moral for bivalence.

> Now what if someone insists that either the statement 'There is an odd perfect number' is true, or else every perfect number is even? He is

justified if he knows of a procedure which will lead him in a finite time either to the determination of a particular odd perfect number or to a general proof that a number assumed to be perfect is even. But if he knows of no such procedure, then he is trying to attach to the statement 'Every perfect number is even' a meaning which lies *beyond* that provided by the training we are given in the use of universal statements; he wants to say [as one might say of] 'Jones was brave', that its truth may lie in a region directly accessible only to God, which human beings can never survey.

(*ibid.*)

Brouwer's intuitionism is based on a Kantian view of our intuition of time while Dummett's is based on a Wittgensteinian account of meaning which leads him to a kind of verificationism. The sources of their respective constructive mathematics are quite different, but the result is the same. We can only assert what we can actually prove by providing explicit examples. Classical logic is not part of our legitimate tool box.

Logic

There are a variety of attitudes towards logic, but most logicians and mathematicians see it as *prior* to and *normative* for mathematics. Logic is a tool for developing mathematics; an instrument for drawing correct inferences. Given any mathematical statement, the rules of logic can be used to make manifest the logical consequences. But in drawing inferences we do *not* create new truths; we merely reveal those that *already* exist. The constructivist attitude is completely different: logic is not prior to mathematics, but comes after; it is not normative, but descriptive. Mathematical constructions can be described in language, and the resulting statements will have a logical structure, but it need not be the structure of classical logic. Let's look at intuitionistic logic in a bit of detail. (Here to some extent I follow Heyting (1956) and use his symbols for the connectives.)

⊦   P   *means* P has been proven (i.e. there is a construction which justifies asserting P).

⊦   P ∨ Q   *means* P has been proven or Q has been proven.

⊦   P ∧ Q   *means* P has been proven and Q has been proven.

⊦   ¬P   *means* there is a construction which deduces a contradiction from the supposed proof of P.

⊦   P → Q *means* there is a construction which when added to the construction of P automatically gives Q.

We can now look at a number of theorems and non-theorems. To show that something is not a theorem we need a counter-example. I'll provide some, but I stress that all the counter-examples to be given are intuitionistic examples; they would not be accepted as genuine counter-examples by anyone who accepts classical logic. An intuitionist, for example, thinks there is no fact of the matter about the existence of a sequence of seven consecutive 7s in the infinite expansion of π. Counter-examples typically play on this sort of consideration. A classical logician would, of course, say that either there is or there is not such a sequence in π; we merely don't know which.

All the propositions to follow are classically valid. Those marked* are *not* intuitionistically valid.

$$\vdash \quad \neg\neg(P \vee \neg P)$$
$$\vdash \quad P \vee \neg P*$$

Counter-example: Define the number n as equal to 1, if there is a sequence of seven 7s in the infinite expansion of π, and equal to 0, if there is no such sequence. Now consider the statement $n = 1$ or $n \neq 1$. In order for this to be true, we need a proof of the existence of seven 7s for $n = 1$, or a proof that there is no such sequence for $n \neq 1$. We have a proof of neither, so we cannot assert the statement $n = 1$ or $n \neq 1$; that is, we cannot assert $P \vee \neg P$.

$$\vdash \quad \neg(P \vee Q) \rightarrow \neg P \wedge \neg Q$$
$$\vdash \quad \neg(P \wedge Q) \rightarrow \neg P \vee \neg Q*$$

Counter-example: Exercise for the reader.

$$\vdash \quad (P \rightarrow Q) \rightarrow (\neg Q \rightarrow \neg P)$$
$$\vdash \quad (\neg Q \rightarrow \neg P) \rightarrow (P \rightarrow Q)*$$

Counter-example: Exercise for the reader.

$$\vdash \quad P \rightarrow \neg\neg P$$
$$\vdash \quad \neg\neg P \rightarrow P*$$

Counter-example: Define the number r as follows: $r = 0.3$, if there is a sequence of seven consecutive 7s in the infinite expansion of π; $r = 0.333 \ldots$ if there is no such sequence. Now assume that r is not a rational number (i.e. assume $\neg P$). Therefore, $r \neq 0.3$ (since 3/10 is certainly a rational number). Thus, there is no sequence of seven 7s in π. Therefore, $r = 0.333 \ldots = 1/3$. But this contradicts the assumption that r is not rational. Thus, the assumption that r is not rational, i.e. $\neg P$, is false, and so we have $\neg\neg P$. Can we infer P from this? No, since to assert P is to assert that r is rational, and we can only do this when we have a proof. To prove that r is rational we need to find integers p and q such that $r = p/q$. But to do this we would need a proof that there is a sequence of seven 7s in π or a proof that there is no such sequence. No proof exists. So we cannot assert P, even though we can assert $\neg\neg$P.

Intuitionistic logic has become an interesting subject in its own right, with important relations to category theory, to forcing in set theory, to computer science, and to many other fields (see Bell and Machover 1977). This brief taste must suffice, however, to bring us to our next topic, the consideration of some problems for the constructive approach to the philosophy of mathematics.

Problems

I'll touch on only a few of the general difficulties of constructive mathematics. One of the most important and effective criticisms of constructive mathematics, especially Brouwer's intuitionism, would focus on its highly problematic view of introspection and the excessively private view of the mind. The heavy emphasis on introspection borders on idealism. In the next chapter I'll turn to this problem briefly, but here I'll avoid further discussion of this difficulty, since it takes us into general issues in philosophy and psychology and away from mathematics.

The finite but very large

The crucial divide for constructivists is between the finite and the infinite; they limit meaningful mathematics to the former. Searching through a finite set is something we can do – in principle. Not so an infinite set. Without a proof, the assertion that all even numbers (greater than two) are equal to the sum of two primes (i.e. Goldbach's conjecture) is neither true nor false, according to Brouwer and company; there is no fact of the matter to be known. The Platonist, by contrast, declares that there is a fact, but pleads ignorance as to which it is. If we restrict Goldbach's conjecture to a finite set, then any constructivist is happy to assert meaningfulness, even if ignorant of what the truth is. Thus, 'All even numbers less than 10^2 are equal to the sum of two primes' is perfectly meaningful. And, of course, it's easy to see why. We can inspect each even number (there are only 50) in the set and determine whether they all are equal to some pair of primes. But what about even numbers less than $10^{1,000,000,000}$? The same procedure for determining the answer works here, too, but only in principle. There is no hope of checking each case even if all humans worked all their lives with all the supercomputers for the entire history of the universe. Nevertheless, it is constructively acceptable.

Brouwer and others don't hesitate to talk about infinite sets, but it is always the 'potential infinite' that they have in mind. A realist typically thinks of the set of numbers as an actual infinite; the numbers are already there and there are infinitely many of them. Champions of the potential infinite say that the set of numbers is infinite only in the sense that however far we count, we can always continue. The numbers aren't already there, but are being created as we count. At any stage in counting, we have only created a finite number. Calling it infinite is simply a way of saying that we won't suddenly be unable to go on.

The distinction between potential and actual infinities was perhaps created by Aristotle in his attempt to overcome Zeno's paradoxes. Suppose I draw a line. How many points are there on it? You're likely to say, An infinite number of points. But Aristotle would say, None. However, Aristotle maintained, if I make a cut mark in the line, I create a point. Now it has one point. I can do this over and over again, creating many points. But, two things must be noted: first, at any stage in this process I have created only a finite number of points on the line, and second, no matter how many points I have created, I can always create more. Thus, says Aristotle, the points on the line are a potential infinity, not an actual infinity.[1]

However, no constructivist wants to stop at the small finite number of things we have actually created. Any finite system, however large, is fair game. Recall the Bishop passage quoted above (p. 117):

> The transcendence of mathematics demands that it should not be confined to computations that I can perform, or you can perform, or 100 men working 100 years with 100 digital computers can perform. Any computation that can be performed by a finite intelligence – any computation that has a finite number of steps – is permissible.
>
> (*loc. cit.*)

But why should it be? Is the infinite any more inaccessible than the very large finite? We might as well hang for a sheep as a lamb – or shouldn't we? I doubt that there is any straightforward answer to this. The situation is similar to one in debates over scientific realism. Van Fraassen (1980), for example, holds that legitimate scientific belief should be confined to the observable; to accept a theory is to believe that it is empirically adequate. This empirical adequacy stretches over *all* space and time, far beyond anything humans could hope to check. Thus, science goes well beyond actual human experience. However, says Van Fraassen, we should not extend our beliefs to include theoretical statements. Beliefs about the observable, like the finite, are something we can and must accept even though it eludes our grasp in many cases. Beliefs about the unobservable, like the infinite, are, according to Van Fraassen, simply not admissible. There are many important disanalogies between Van Fraassen and the constructivists, but this one holds. In the case of Van Fraassen vs. the scientific realists, the debate, in the eyes of many, is at a stand-off; neither side has the advantage on this point. Given the similarity between the two cases, neither the constructivists nor their realist opponents are likely to gain a clear upper hand in the parallel question.

The recent solution of the so-called *n*-body problem provides an interesting example for constructivists. The previous example of Goldbach's conjecture is from pure mathematics; this one is intimately connected to physics. The *n*-body problem arose in Newtonian celestial mechanics. Given the positions and momenta of *n* bodies interacting under Newton's laws (but not colliding), what are their positions and momenta after any given temporal interval? (Given the positions, velocities, and masses of the sun and the planets today, for instance, what will they be in a hundred years?) The special case of two bodies was solved long ago by one of the Bernoullis. The general problem was set by Mittag-Leffler, Hermite and Weierstrass in a famous prize competition in the 1890s. They were looking for a convergent power-series solution. Poincaré (the winner of the contest) gave an impossibility proof, of a sort; but that was confined to a specific method of solving the problem. It still left open the possibility of finding a solution, and that was recently done (Q. Wang 1991).

The solution is indeed a series solution which meets all the expectations of the original posers of the problem. And yet, remarkable though the achievement is, there is a sense in which nothing has changed. The problem is that the series converges *very* slowly. One has to compute millions of terms in the series before getting a reasonable approximation to the actual value for even a very short time ahead. From either a practical or a theoretical point of view the series solution is of no value – though it fully satisfies Brouwer's or Bishop's or Dummett's demands for an existence proof.

If I may be allowed an autobiographical remark, I find this sort of objection much more convincing now than I did a few years ago, and I suspect that I'm not alone. The difference comes, I think, from experience with a PC. Like anyone with a computer, it matters little to me what my machine can do *in principle*. I care about what it can do *quickly* and I have become sensitive to the efficiency of various different software programmes and the relative difficulty of the problems that I set. I know that *Maple* or *Mathematica* can factor 'small' numbers quickly, but if I give it something with, say, 100 digits, then my PC will be working on it for longer than I care to wait around. The 'in principle' consideration plays no part in my practical life. This feasibility problem is simply not an issue for constructivists, for whom the main dichotomy is finite–infinite. They are insensitive to the real dichotomy for working mathematicians or PC users: feasible vs. infeasible. One wants to cry out: Give us the power of Platonism or give us computational practicality. Constructivists give us neither.

Applied mathematics

The n-body problem arises in physics, but can be considered from a pure mathematics point of view. The role of applied mathematics per se is not much discussed by constructivists; the philosophical debate has focused on pure mathematics almost exclusively. Herman Weyl went through a constructive (or rather 'semi-intuitionist') phase (Weyl 1918), but later in his career he found that applied mathematics decided the issue: if constructive mathematics can't do justice to the sciences, then it should be tossed out and classical mathematics should be retained (Weyl 1949).

Most applied mathematics is not only constructive, it is finite. Even when we use infinite mathematics in physics, it is not the physics that forces it upon us. We use real numbers, for example, to measure distances. But we can never measure with perfect accuracy; our measures are all to some distance, $\pm$ a bit. This means that rational numbers are perfectly adequate for measuring. However, applied mathematics is often much fancier than this.

In an earlier chapter I discussed Hartry Field's view that science without mathematics is possible. He holds this view in opposition to Quine and Putnam who say mathematics is indispensable to science. Let us not worry about who is right in this debate, but only ask the more restricted question: Is all the mathematics that science actually does use constructive? It turns out that the answer is very likely No.

The mathematics used by quantum mechanics consists of Hilbert spaces, which are linear or vector spaces of (possibly) infinite dimension. States (e.g. the state of an electron) are represented by a vector ψ in the Hilbert space and properties (e.g. position, momentum, spin, or energy of the electron) by linear operators. These are functions with special properties that are defined on vectors such as ψ. Recent interest in chaos and complexity, etc. has led to some significant research into the properties of linear operators. Pour-El and Richards (1983) have proven that a class of linear operators which includes some typically used in quantum mechanics do not preserve certain computability conditions. This means that they are non-constructive functions. From a constructive point of view they are simply not legitimate. But they are crucial for the Hilbert space apparatus of standard quantum mechanics.

A second example is Gleason's theorem which concerns measures on the closed subspaces of a Hilbert space. It is of central importance to the foundations of quantum mechanics since it rules out a wide class of hidden variable theories; so, the theorem is central to our understanding of how nature works. However, no constructive proof of this theorem is possible.

These examples are too complicated to be described here, but the moral is important. As Hellman[2] and others have pointed out, the conclusion seems inevitable: constructive mathematics is not rich enough to serve the needs of science.

Most participants in the debate over constructive mathematics agree on this: If science needs classical mathematics, then that effectively torpedoes constructive approaches. What they disagree on is whether there is any science that can't be properly served by constructive mathematics. The claims and counter-claims have turned on sophisticated examples. It seems to me, however, that the shortcomings of constructive mathematics can be displayed easily with less esoteric cases, which is what I now aim to do. Consider the intermediate value theorem (described in Chapter 3):

> If f is continuous on the interval $[a, b]$ and there is a C between $f(a)$ and $f(b)$, then there is a c between a and b such that $f(c) = C$.

This is a classical theorem, of central importance in analysis, but not constructively valid. Its proof relies on the law of trichotomy (i.e. for every x and y in $\mathbf{R}$: $x < y$, or $x = y$, or $x > y$), and the existence of a least upper bound (i.e. any set of real numbers with an upper bound has a least upper bound), neither of which is constructively provable.

I'll set up a simple case in physics that requires the theorem. Let us suppose a substance changes its temperature *continuously* (though perhaps not monotonically) from T_1 at time t_1 to T_2 at time t_2. Let us further suppose that there is a temperature sensitive device inserted in the substance such that if it registers *exactly* temperature T, where $T_1 < T < T_2$, it will send a signal to a remote location that will detonate a bomb.

A physicist using classical mathematics will make the following inference: Since the substance changed temperature continuously from T_1 to T_2 and T is between these two values, there was a time t between t_1 and t_2 when the temperature was T. Thus the remote bomb exploded.

A physicist using constructive mathematics cannot make this inference. She cannot infer that there was a time t when the temperature was T, and so she cannot conclude that the remote bomb did indeed explode. The explosion is an independent fact that will have to be empirically checked on its own.

Measuring the temperature *exactly* is crucial, since a constructive version of the theorem allows us to approach the intermediate value with arbitrary closeness, though never achieve it. A device that is sensitive to an exact temperature is an idealized case, like a frictionless plane. There may be no such device in the world, but physics still deals with them in thought experiments and mathematics must apply to them, as well. The history of physics without imaginary examples is unthinkable, and a mathematical theory that cannot do justice to thought experiments is as problematic as one that cannot do justice to real experiments. (For more on thought experiments in the natural sciences, see Brown 1991.) Of course, we cannot run actual experiments to see if the predicted explosions actually happen, because we cannot set up devices sensitive to exact temperatures. Nevertheless, the cost of abandoning classical mathematics is apparent. The crucial difference has to do with organizational ability. Constructive mathematics allows less systematization and predictive power in science than does its classical rival. The inability of a physicist using constructive mathematics to draw relevant inferences if only in thought experiments must be seen as a sizeable obstacle to taking constructive mathematics seriously in general.

Negation

What sense can we make of negation in constructive mathematics? This is a special problem for intuitionism, since Brouwer emphasizes that it is a *languageless* activity. Heyting characterizes a negative proposition this way: '⊢ ¬P can be asserted if and only if we possess a construction which from the supposition that a construction P were carried out, leads to a contradiction' (1956: 102). His larger gloss is worth quoting at length:

> Every mathematical assertion can be expressed in the form: 'I have effected the construction A in my mind.' The mathematical negation of this assertion can be expressed as 'I have effected in my mind a construction B, which deduces a contradiction from the supposition that the construction A were brought to an end', which is again of the same form. On the contrary, the factual negation of the first assertion is: 'I have not effected the construction A in my mind'; this statement has not the form of a mathematical assertion.
>
> (Heyting 1956: 19)

So, ¬P is not the claim that P isn't constructed; but rather, is supposed to describe a construction itself. Yet, what sort of construction could ¬P be? And what is 'the supposition that a construction P were carried out', if it is not a linguistic entity? Since P is absurd, it can't be an actual construction, but it could be a proposition (sentence, statement) that P has been constructed. And finally, what is the 'contradiction' that Heyting refers to, if not a linguistic entity? (For intuitionist purposes, proposition, statement and sentence would all be part of language.) I can say in words that I've baked a cake that is both round and square, but it is certainly not an actual cake.

If we are to take the languageless nature of mathematics seriously, as Brouwer would urge us to do, then the notion of negation may have to go. Suppositions and contradictions are in language – not in reality, whether mathematical reality is constructed or not. Intuitionism is thought already to abandon too much of classical mathematics. Without negation, even more will go out the window, leaving a very impoverished residue.

Arithmetic versus geometry

Brouwer, as I mentioned above, accepts the Kantian account of arithmetic, based on the perception of time. Geometry is a completely different matter for him. Unlike synthetic a priori arithmetic, Brouwer holds the anti-Kantian view that geometry is analytic. Given these two quite distinct attitudes to the two different branches of mathematics, some sort of tension must seem inevitable when we consider those mathematical theories such as knot theory or graph theory which make essential use of different forms of representation. (Recall the discussion of this point in the previous two chapters.) Graph theory, for example, uses both a (finite) set-theoretic representation and a diagrammatic representation. These are, respectively, synthetic a priori and analytic, for Brouwer. At best constructivist claims hold for the set-theoretic aspects of graph theory; those that stem from graph diagrams do not reduce to computations with numbers.

Exhibiting an instance

A standard constructivist claim is this: A legitimate proof gives a method of constructing explicit instances. Recall Bishop's injunction: 'When a man proves a positive integer exists, he should show how to find it' (*loc. cit.*). Dummett asserted the same point: 'We are thus justified in asserting that there is a number with a certain property only if we have a method for finding a particular number with that property' (*loc. cit.*). This is somewhat problematic, perhaps even false. There may well be constructively acceptable proofs that don't give us an example.

Consider the question: Is 53,461 prime or composite? (Remember, a composite number is the product of primes.) There is a constructively valid theorem of Fermat that says: *Any prime of the form 4n + 1 can be expressed as the sum of two perfect squares in one and only one way* (e.g. $13 = 4(3) + 1 = 9 + 4 = 3^2 + 2^2$). Examining 53,461 we see

$$53,461 = 4(13,365) + 1 = 10^2 + 231^2 = 105^2 + 206^2$$

Thus, it has the form $4n + 1$ and yet it can be written as the sum of two squares in *two* different ways. So, by Fermat's theorem it cannot be prime, but must instead be composite.

However, in spite of constructively proving that it is composite, no actual factors are given. And no way of finding them is indicated by the proof. Of course, there is a different constructive method (known as the Sieve of Erathosenes) which – in principle – would reveal the prime factors; but it's not practical and – most importantly – it does not stem from the constructive proof presented here that the number is composite. The example given was a bit contrived (Dunham 1994), but for a truly huge number, we might establish its composite nature by means of Fermat's conditions, without any humanly practical hope of ever finding prime factors. Thus, we would have a constructively valid proof that prime factors exist, *without* being able actually to exhibit any.[3]

It's an alleged virtue, often repeated, that constructive proofs give us more information – they give us actual methods of producing instances. This is not always true. We have seen this fail in two ways. Above, the series solution of the n-body problem provided a constructive proof, but the convergence of the series is so slow that, from a practical point of view, we really don't have any instance at all; there is nothing here that we can point to and use. In the example just given concerning composite numbers, we have a constructive proof and we can – in principle – find prime factors. But the constructive proof does *not* provide those prime factors. It is a different, very impractical technique that must be used to find the actual instances.

The loss of many classical results

The biggest complaint against constructive mathematics is that accepting this way of doing things would result in the loss of much of classical mathematics. I'll illustrate with an example, the Bolzano–Weierstrass theorem which says: *Every bounded infinite set S has at least one cluster point.* (A point x is a *cluster point*, or a *limit point*, if and only if every neighbourhood of x has members in S; otherwise x is an *isolated point*.) Pictures (Figure 8.1) are a help here (but only as an aid, not as a proof).

To prove this theorem we assume that S is a bounded set with infinitely many points (Figure 8.1(a)). To keep things simple, we will assume that the points are all on the plane. Since S is bounded, it can be contained in a square, Q_0, which has sides of length L (Figure 8.1(b)).

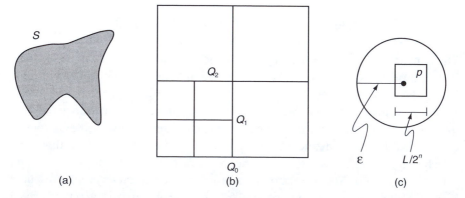

(a) (b) (c)

Figure 8.1 The Bolzano–Weierstrass theorem

Now we divide Q_0 into quarters. At least one of these smaller squares, say Q_1, will have infinitely many points of S. Next, we divide Q_1 into four, and again, one of these, say Q_2, will have infinitely many points of S. We continue in this way forming the sequence: $Q_0, Q_1, Q_2, \ldots$, which is an infinite sequence of subsets: $\ldots \subset Q_3 \subset Q_2 \subset Q_1 \subset Q_0$. Each has infinitely many points of S. By the nested set property, there is at least one point, p, common to them all.

Consider next an arbitrary neighbourhood around p (Figure 8.1(c)). By making the number of divisions large enough, the square Q_n, with sides of length $L/2^n$ will be contained within the radius ε. Thus, p is a cluster point. This ends the proof.

The Bolzano–Weierstrass theorem is central to classical analysis. Current real and complex analysis, current topology, and many other areas of contemporary mathematics would be much diminished without it. But it is not constructively legitimate, since it proves the *existence* of the cluster point p, but gives no method whatsoever for actually finding or constructing p. This is the sort of theorem that would be lost, if constructive mathematics came to prevail – a price too high for most to pay.

Further Reading

Many of the Brouwer's influential works can be found in translation in various anthologies, such as Benacerraf and Putnam (eds), *Readings in the Philosophy of Mathematics*; Ewald (ed.), *From Kant to Hilbert*; and Mancosu (ed.), *From Brouwer to Hilbert*. Heyting's *Intuitionism: An Introduction* is a good place to start and so is Dummett's *Introduction to Intuitionism*. For mathematically advanced readers, Bishop and Bridges, *Constructive Mathematics* is the best source. It begins with Bishop's famous 'Manifesto', a clear statement of his philosophical views on the nature of mathematics.

CHAPTER 9
Proofs, Pictures and Procedures in Wittgenstein

A Picture and a Problem

It's a curious fact about pictures that most philosophers never use them. Browse through the works of the great philosophers and you will find almost no pictures at all. Wittgenstein is an exception; his works are filled with little sketches and rough diagrams. Descartes is another, though his diagrams are mostly connected with the scientific aspects of his work. If we leave Descartes aside, it's probably no exaggeration to say that there are as many pictures in Wittgenstein's published works as there are in all the other great philosophers combined. And what he has to say about pictures, diagrams and illustrations is interesting and important – though often only implicit in his remarks. Wittgenstein in later years may have abandoned his 'picture theory of meaning', but not his fondness for pictures.

Wittgenstein often cites some picture as proving this or that result. And at one point in the *Remarks on the Foundations of Mathematics* he declares, 'mathematics is a motley of techniques of proof' (*RFM*, III-46). The suggestion is that there are not only verbal/symbolic proofs in mathematics, but there are picture-proofs, too. This seems to me to be exactly right. In fact, I'm tempted to say further that mathematics is so rich that there are indefinitely many ways to prove anything – verbal/symbolic derivations and pictures are just two. At any rate, Wittgenstein certainly thought that pictures are included in the 'motley' and this is the topic that I want to explore. Here (Figure 9.1) is one of his examples along with his discussion:

Consider a mechanism. For example this one:

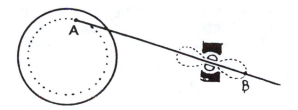

Figure 9.1

While the point *A* describes a circle, *B* describes a figure eight. Now we write this down as a proposition of kinematics.

When I work the mechanism its movement proves the proposition to me; as would a construction on paper. The proposition corresponds e.g. to a picture of the mechanism with the paths of the points *A* and *B* drawn in. Thus it is in a certain respect a picture of that movement. It holds fast what the *proof* shews me. Or – what it persuades me of.

If the proof registers the procedure according to the rule, then by doing this it produces a new concept.

In producing a new concept it convinces me of something. For it is essential to this conviction that the procedure according to these rules must always produce the same configuration. ('Same', that is, by our ordinary rules of comparison and copying.)

With this is connected the fact that we can say that proof must shew the existence of an internal relation. For the internal relation is the operation producing one structure from another, seen as equivalent to the picture of the transition itself – so that now the transition according to this series of configurations is *eo ipso* a transition according to those rules for operating.

In producing a concept, the proof convinces me of something: what it convinces me of is expressed in the proposition that it has proved. . . .

The picture (proof-picture) is an instrument producing conviction.

(*RFM*, VII-72)

Wittgenstein's picture does indeed convince us of the truth of this proposition:

As the point A moves in a circle, the point B moves in a figure eight.

Wittgenstein is quite right to lay such stress on diagrams, often in the face of conventional mathematical wisdom. However, there is something remarkable about this particular diagram, something quite instructive. It's faulty. Yet it works perfectly well at helping us grasp what's going on.

The diagram is mis-drawn in two key respects. (You might try to detect them before reading on.)

For one thing, *A* is midway between its leftmost and its rightmost positions. So *B* should be midway as well. Instead of being drawn at its rightmost position, *B* should be drawn in the middle of the mechanism through which it slides. Second, the axis of symmetry of the dotted figure-eight should be horizontal, not drawn as lying along the moving rod.

These are serious flaws in the diagram. And yet it doesn't matter. We are convinced of the proposition anyway. But what we are probably convinced of is that the motion of *B* is *a* figure eight, not that it is precisely *the* figure eight as drawn. Here is the diagram drawn properly (Figure 9.2).

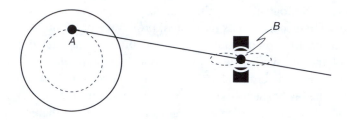

Figure. 9.2 Wittgenstein's picture correctly drawn: the figure eight is horizontal and *B*
is at the centre of the sliding mechanism

Let me digress for a moment and talk about the origin of the faulty picture. G.H. von Wright, one of the editors of Wittgenstein's posthumous *Remarks on the Foundations of Mathematics*, tells me (in a private communication) that the published drawing was provided by the editors, and was based on Wittgenstein's own rough drawing in the ms. Von Wright agrees that both mistakes are in the published version. In Wittgenstein's own (a photocopy of which von Wright kindly sent me) things are not so clear. Von Wright thinks one mistake is definite (the incorrect position of B) but that the figure eight is correctly oriented. I'm less confident, and think that the artist who copied Wittgenstein's rough picture could quite reasonably assume Wittgenstein intended the figure eight oriented along the rod instead of horizontally.

Of course, it doesn't really matter what the origin of the mistake is. The philosophical problem arises, regardless. Our problem (and Wittgenstein's) is this: Why does the diagram work, in spite of its flaws? I will try to answer this question as part of an exposition and critical response to some main themes in Wittgenstein's philosophy of mathematics.

Following a Rule

People *use* mathematics. This is an important fact never to be lost sight of, though it may seem a trivial observation. Most accounts of mathematics focus on mathematics itself. Only as a kind of afterthought is the fact taken up that people apply mathematics in a variety of everyday situations. This attitude is perfectly understandable, especially if we think of mathematical results as a body of truths. After all, our account of the physical sciences is unaffected by the fact that people do use in a practical way some of those sciences (such as mechanics), but don't use others at all (e.g. astrophysics). Usability seems largely irrelevant to believability. Platonism is an extreme example of this attitude. It includes a well-developed and, arguably, a highly successful account of the nature of mathematics. Its (alleged) problems stem from the apparent complete disconnection of embodied creatures such as we are from the mathematical realm; so, it's a mystery how we are able to use it.

Wittgenstein reverses this. He focuses on how people use mathematics, especially in simple situations of counting and measuring. It is only after sorting through this and resolving a sceptical paradox of his own making that he then comes (in some sense) to general conclusions about the nature of mathematics. I insert the caveat 'in some sense' because Wittgenstein repeatedly claimed to be not proposing any theories. In some respects this is an importantly true self-description. But the plain fact is that he often did, so I shall not be too concerned with his infamous claim not to be theorizing.

One of Wittgenstein's main concerns is with the idea of *following a rule*. How do we know someone is indeed obeying a particular rule? Joe Blow drives at 100 k.p.h. in a 50-k.p.h. zone. The police officer who pulls him over scoffs at hearing that his speed was quite within the rules of the road and gives him a ticket. Mary Doe earns a large salary, has no legitimate deductions, yet she pays not a penny in tax. The revenue people ridicule her claim that she has scrupulously obeyed the tax laws, and they impose a steep penalty. Tommy Twaddle adds 2,000 to 3,000 and gets 23,000,000 for the answer; he claims to be following the rule for addition. Should we scoff and send him back to remedial arithmetic?

Of course, rules can be vague – but not here. We feel perfectly secure in saying that the traffic rules, the tax rules and the mathematical rules were each flagrantly violated. But should we be so sure? Wittgenstein would not join us in thinking things quite so obvious. The very idea of 'following a rule' is, he thinks, highly problematic. It is one of the central preoccupations of his later philosophical work, with implications for language and mind as well as for our understanding of mathematics.

Suppose a teacher (she) lists a few numbers in sequence and asks a student (he) to continue:

1, 4, 9, 16, . . .

What is the next number in the sequence? The student will likely answer 25. Why? Probably he will reason that the sequence obeys the rule of taking each number in turn and squaring it, i.e. 1^2, 2^2, 3^2, 4^2, and so on; the next number in the sequence would then be 5^2 which is 25.

But what if he answered 27? Would this be wrong? His reasoning might be as follows: The sequence is growing by adding successive odd prime numbers to the elements of the sequence. The odd primes are: 3, 5, 7, 11, 13, 17, . . . Thus the sequence is:

$$1$$
$$4 = 1 + 3$$
$$9 = 4 + 5$$
$$16 = 9 + 7$$
$$\vdots$$

And so the next number in the sequence is $16 + 11$ which equals 27.

Is either answer right? Perhaps a better question is: In what sense could an answer be *wrong*? Both answers were based on grasping a rule, then continuing that particular rule correctly. Both rules are compatible with the first four numbers in the sequence and, of course, compatible with each other to that extent. (Indeed, there are infinitely many different sequences which overlap at any finite number of initial elements.) Neither answer can be faulted in either of these two regards. There is nothing intrinsic to the answers which make either of them wrong.

The only sense in which an answer could be wrong has to do with the questioner's *intentions*. If the teacher had intended the sequence based on squares rather than on adding primes, then 27 is the wrong answer, 25 is right. We can assume that both teacher and student readily understand the difference between these two sequences, which are generated by different rules, and that both have often in the past calculated such numbers as the square of 5 and the sum of 16 and 11. So there is no problem about what would be the next number in each of these respective sequences. Moreover, there seems to be no problem about the notion of the questioner's intentions; had *he* known which sequence *she* intended, *he* could have given the right answer.

Now let us suppose that the questioner did indeed have the sequence in mind which is generated by the rule of squaring the successive numbers and is in the process of teaching the student the operation of squaring a number. The teacher lists a few numbers in the sequence: 1, 4, 9, 16, 25, . . .; she shows how these come from the operation of squaring, then she asks the student to 'go on in the same way.' He continues the sequence: . . ., 36, 49, 64, . . ., and she is quite happy with his progress.

At this point we will further suppose that in all of human history no one has ever squared a number greater than 100. (For some much larger, but still finite number, this will be true; but a small number will, of course, illustrate more easily.) The student continues:

$$\vdots$$
$$98^2 = 9{,}604$$
$$99^2 = 9{,}801$$
$$100^2 = 10{,}000$$
$$101^2 = 11$$
$$102^2 = 12$$
$$\vdots$$

The teacher balks at this; she says that the answers after the square of 100 are wrong, and she claims that he has not followed the rule properly. The great majority of us would agree with her.[1] (At least we would likely agree after calculating; remember we are squaring numbers this size for the first time in human history.) In response the student recalculates several times, always getting the same answers as he got earlier, so he becomes convinced that no computational mistake was made. He also claims to have followed the rule that was given by the teacher and to have carried on as instructed, 'in the same way'.

So, why is the student's answer wrong? And are we even sure that it is wrong? Perhaps it isn't. To make matters worse, we may suppose that our student is an emissary from a postmodern literary department. His car sports the bumper sticker 'Everything is text', to which he cheerfully adds 'And an author's intentions are just more text'. In the light of this he might maintain that his answer is right, that is, his answer is *just as correct as any other*. In particular, he denies that there is anything that the teacher said, or did, or had in mind, that determines what the 'true' answer is.

The argument would seem to be time-symmetric. Suppose, to change the example slightly, I am teaching myself. I tell myself on Tuesday to 'go on in the same way'. Then, on Wednesday, I raise sceptical doubts about what I could have meant on the previous day. If I can claim today that I didn't know what I meant by such and such yesterday, then tomorrow when I attempt to 'go on in the same way', I will be able to say I didn't know what I meant today. Consequently, right now, though I am trying my best, I don't know what I mean by such and such.

Is there anything that the teacher might do to justify her answer as the correct one and any other answer as false? Is there any response to this extreme scepticism – a scepticism that Kripke (1982) calls the most profound in all philosophy to date?

Let's canvass some explanations. In looking at these we'll begin to see the depth of Wittgenstein's problem and why there is some plausibility to this extreme scepticism. There are three things to be interested in here: first, Wittgenstein's *actual* account of the matter; second, the account he *ought* to have offered, given his other views and general framework; and third, the *right* account. They might not be the same – indeed, it would be surprising if they were. Of these three, it is the second – what Wittgenstein *ought* to have said given his other views – that I will mainly focus on. There is a large and difficult literature devoted to figuring out just what Wittgenstein actually thinks. (In large part the problem is due to the obscurity of Wittgenstein's presentation.) So far, less has been written on what the absolutely right answer might be and virtually nothing at all has been said about what the right answer within a Wittgenstein framework should be. Kripke (1982), for example, contents himself with saying what Wittgenstein's argument actually is. (The expository parts of this chapter have been much influenced by Kripke's work.) The depth of Wittgenstein's rule-following problem for any approach to the philosophy of mathematics has only gradually become apparent. So even if we reject Wittgenstein's particular solution, the problem remains to be dealt with.

Platonism

Recall the Platonism of Frege, who distinguishes between (i) our ideas (which are psychological entities); (ii) 'thoughts', as he calls them (which are the content of our ideas); and (iii) the sentences we use to express them (which are 'things of the outer world', physical entities such as ink marks or sound waves).

> Thus the thought, for example, which we express in the Pythagorean theorem is timelessly true, true independently of whether anyone takes it to be true. It needs no bearer. It is not true for the first time when it is discovered, but is like a planet which, already before anyone has seen it, has been in interaction with other planets.
>
> (Frege 1918: 523)

In keeping with this sentiment, we can return to our sequence and take it (or any function) to be an independently existing entity, a set containing infinitely many ordered pairs: $S = \{\langle 1,1 \rangle, \langle 2,4 \rangle, \langle 3,9 \rangle, \ldots, \langle 99,9801 \rangle, \langle 100,10000 \rangle, \langle 101,10201 \rangle, \langle 102,10404 \rangle, \ldots\}$. The ordered pairs $\langle 101,11 \rangle$ and $\langle 102,12 \rangle$ are not members of this set. And that, very simply, is why the student's answer was wrong.

The crucial assumption in this account – both natural and plausible – is that there are mathematical facts just waiting for us to discover. Those who grasp the squaring function answer correctly, even when faced with particular examples they have never before encountered. The student simply didn't grasp the function, or perhaps grasped some other.

Platonism was just one of Wittgenstein's many targets. He was quite dismissive, of course, but it seemed to him rather obviously wrong and hardly in need of detailed refutation. Typical of his dismissive remarks is the rhetorical question, 'Is it already mathematical alchemy, that mathematical propositions are regarded as statements about mathematical objects, – and mathematics as the exploration of objects?' (*RFM*, V-16). When Wittgenstein needed an example of the view he so disliked, he often chose G.H. Hardy, his Cambridge colleague, whom I mentioned earlier. Wittgenstein often refers to Hardy's classic essay, 'Mathematical Proof', which contains such Platonistic pronouncements as this: 'I have myself always thought of a mathematician as in the first instance an *observer*, a man who gazes at a distant range of mountains and notes down his observations' (Hardy 1929: 18).

In response to this type of view, Wittgenstein asserts: 'The mathematician is an inventor, not a discoverer' (*RFM*, I-168). But rather than try to meet Platonists such as Hardy head-on, he remarks 'All that I can do, is shew an easy escape from this obscurity and this glitter of the concepts' (*RFM*, V-16).

Clearly, Wittgenstein will have neither truck nor trade with Platonism. But should we join him? We may be compelled. Let's think again about the sequence discussed above and about the question, What is the next number? Even if the Platonistic account is right about the independent existence of the squaring function with its infinitely many ordered pairs, this still may not help us to understand why one answer is right and all others wrong. Let us assume that the sequence determined by the squaring function actually exists in the Platonic sense. We can still wonder if that is what the teacher had intended. There are, after all, infinitely many functions with the same initial members. Did the teacher intend

$$S = \{\langle 1,1 \rangle, \langle 2,4 \rangle, \langle 3,9 \rangle, \ldots, \langle 99,9801 \rangle, \langle 100,10000 \rangle, \langle 101,10201 \rangle, \ldots\}$$

or did she intend

$$S' = \{\langle 1,1 \rangle, \langle 2,4 \rangle, \langle 3,9 \rangle, \ldots, \langle 99,9801 \rangle, \langle 100,10000 \rangle, \langle 101,11 \rangle, \ldots \}?$$

The problem does not turn on the independent existence of S and S', but rather on the intention: Was it S, or was it S', or yet some other sequence that the teacher had meant? The teacher teaches by giving examples; but the examples cited are exactly the same for the different sequences. Note that this does not involve scepticism about the mathematical entities themselves – the square of 101 really is 10201 – but rather scepticism about which sequence the questioner had in mind. Remember, both teacher and student are computing 101^2 for the first time in human history, so it's not a question of the teacher already having the answer 10201 in mind. The problem is this: Is there anything that the teacher said, did or intended, that *determines* that the right answer is 10201? Our sceptical student claims No. Even if various sequences exist independently of us, just as Platonists claim, still there might be nothing the teacher said, or did, or had in mind, that will pick out a unique sequence. Even if Platonism is true (which Wittgenstein would hotly deny), there is still the problem of knowing if there even is a *right* answer to the question, What's the next element of the sequence?

Algorithms

We might think that no one learns how to compute by merely seeing a few examples. Instead, one learns an algorithm. And the algorithm covers all infinitely many cases, not just the few used to illustrate it. So instead of teaching the operation of squaring with a finite number of instances, why not just state the algorithm explicitly and be done with it?

Unfortunately, this proposal simply pushes the problem back a step. In stating an algorithm, various rules, functions and operations are employed. The initial scepticism about whether S or S′ is intended, can be extended to the rules, functions and operations within the algorithm itself. We are no better off.

Dispositions

Perhaps the teacher had a disposition to give 10201 as the next element in the sequence. It's the disposition that makes this the right answer. (The student, of course, had a different disposition, one which would give 11 as the next element.) Thus, it is not a matter of anything the teacher said, did or intended that determined a unique answer – it was a disposition. The teacher was in a specific dispositional state. Different dispositions would pick out different sequences. And there is an objective matter of fact as to what the disposition was; so there is an objective answer to the question, What is the next number?

There are at least three problems with this attempt to overcome the sceptical student's challenge. For one thing, we would need to have infinitely many dispositions to cover the infinitely many elements of each of the infinitely many different sequences. This seems highly implausible.

Second, dispositional facts seem to rest on categorical facts. Glass, for example, has the disposition to break if hit sharply by a rock. But there is a reason for this, having to do with the molecular structure of glass. Thus the disposition-to-break facts rest on the non-dispositional facts about molecular structure. Similarly, mathematical dispositions would rest on categorical facts about what is said, done, intended, and so on, the very kind of thing that is being challenged by the sceptical student.[2]

Finally, even if one did have a disposition to say '10201', why does this make it the *right* answer? Having a disposition is not the same as having a justification. The teacher claims that 10201 is the right way to continue the sequence;

but having a disposition to say '10201' does not in the least show that 10201 is correct. Meaning and intention have a normative ingredient which tells us how we ought to act. Dispositions (as naturalistic entities) do not capture this at all. (See Kripke (1982) for a development of this important point.) I suspect that the problem here with dispositions is but a special case of the general problem of attempting a naturalistic account of norms.

Knowing Our Own Intentions

Perhaps the teacher is having trouble conveying which function she means, but at least she knows which one she intends. Perhaps nothing she has said nor any action she has taken in the past will help us to pick out a unique sequence; but, she herself knows which one she means, because she knows her own mind.

Wittgenstein's explicit response to this seems right on target. 'But in this case [she has] no criterion of correctness. One would like to say: whatever is going to seem right [to her] is right. And that only means that here we can't talk about "right"' (*PI* §258).

This suggests a comparison with the intuitionists.

Brouwer's Beetle

Wittgenstein has much in common with Brouwer and the intuitionists, especially a finitistic attitude. But on one point they couldn't be further apart. Brouwer's mental constructions are as private as anything could be; indeed, they are almost solipsistic. Wittgenstein, on the other hand, holds that mathematical calculations are essentially public, based on a social practice. He almost identifies the finite with the overtly public: 'Finitism and Behaviourism are quite similar trends. Both say, but surely, all we have here is . . . Both deny the existence of something, both with a view to escaping from a confusion' (*RFM*, II-62). In a particularly striking passage, Wittgenstein says:

> Suppose everyone had a box with something in it: we call it a 'beetle'. No one can look into anyone else's box, and everyone says he knows what a beetle is only by looking at *his* beetle. – Here it would be quite possible for everyone to have something different in his box. One might even imagine such a thing constantly changing. – But suppose the word 'beetle' had a use in these people's language? – If so it would not be used as the name of a thing. The thing in the box has no place in

the language-game at all; not even as a *something*: for the box might even be empty.

(*PI* §293)

There is no room for Brouwer's *private* mental constructions in Wittgenstein's very *public* account of mathematical activity.

Radical Conventionalism

Many truths rest on linguistic conventions. 'Bachelors are unmarried males', for example, rests on the meanings of the words involved, while the truth of 'John is a bachelor' depends on non-linguistic facts, facts concerning John and whether or not he has been through a marriage ceremony. These are often called, respectively, *analytic* and *synthetic* sentences. A one-time popular view had the truths of mathematics and logic rest on such linguistic conventions. The truth of '5 + 7 = 12', according to A.J. Ayer (1936), rests on our conventions concerning the terms '5', '+', '=', and so on. And when mathematicians make new discoveries, they are merely uncovering the consequences of earlier stipulations.

Perhaps the greatest problem with conventionalism of this sort concerns the notion of 'consequence' (see Quine 1936). Often consequences are quite unexpected. They cannot be laid down as explicit conventions in advance, since there are infinitely many of them. So it would seem that conventionalism has to make the kind of Platonistic assumptions about consequences that it so desperately wants to avoid – the consequences are 'already there' just waiting to be discovered.

Wittgenstein has often been read as a conventionalist, but of a radical sort in which even the consequences of our conventions are themselves more conventions. Thus, 'going on in the same way' is not the following of a predetermined pattern; it is not merely the unpacking of the existing consequences, but is instead the laying down of more new conventions. 'However queer it sounds', says Wittgenstein, 'the further expansion of an irrational number is a further expansion of mathematics' (*RFM*, V-9).

In terms of our earlier example, the student is thus laying down a new rule. He is not applying an old rule in a new case, but actually creating a new procedure. The disagreement with the teacher is a fight over which new convention to adopt.

By being a complete conventionalist, Wittgenstein gets around the problem of independently existing consequences. It's one of those not altogether rare cases where the extreme view is more coherent and plausible than the moderate one.

Nevertheless, it may still be wrong. Road builders, for example, might fight over whether to extend the road to the east or to the west. Neither is 'correct', though one way might be more useful than the other. However – and this is the crucial point – it is possible to actually do both, that is, to build two extensions in different directions. They can happily coexist. Why can't mathematics – if it's just a body of conventions – be developed, similarly, in different directions simultaneously? There would seem to be no objection on Wittgensteinian grounds. And yet it is not done. There is only one 'addition', only one 'multiplication', etc. They are unique. Wittgenstein would need to explain this uniqueness in mathematics that does not exist in road building. Is uniqueness in mathematics itself just another convention? The kinds of analogies that spring to mind to explain radical conventionalism tend to dissolve when inspected more closely.

Dummett finds it not credible that someone could understand all the concepts in a proof yet still reject that proof. He notes that while Wittgenstein claims this could happen, he offers no plausible instances. 'The examples given in Wittgenstein's book are – amazingly for him – thin and unconvincing. I think that this is a fairly sure sign that there is something wrong with Wittgenstein's account' (Dummett 1959a: 430).

Bizarre Examples

Wittgenstein often discusses bizarre behaviours and talks about them in terms of different 'forms of life'. These problems should not be confused with our problem, though commentators, and even Wittgenstein himself, often run these things together.

In one of his examples, Wittgenstein has us imagine people who buy and sell lumber in odd ways. The price is fixed, say, by the area that a pile of lumber covers. Spread the pile out and the price goes up. We might try to teach them that this is silly by making the pile higher on a much smaller base. But in response they just say that now the cost is much less, and so on. This is indeed odd behaviour, but it has nothing to do with mathematics, even though elaborate computations might be made to set a price for the lumber. I will use Platonistic language to argue this because it is, first of all, easier to do so and, second, it will show how untouched Platonism is by these sorts of bizarre examples.

The mathematical realm contains all possible structures. Any way the world could be is isomorphic (or at least homomorphic) to some mathematical structure. Different ways of applying mathematics to the world amount to associating the world (or some relevant part of the world) with different mathematical structures (see Chapter 4).

Wittgenstein's fictitious examples are no more bizarre than some real examples chosen from modern physics. Consider a simple case like the addition of velocities. A person at the back of a plane flying at 800 k.p.h. throws a ball forward at 100 k.p.h. What is the speed of the ball with respect to the ground? Common sense and classical physics both say 900 k.p.h. But those who have adopted special relativity would say this is not so, that the correct answer is actually a bit less.

They disagree, but do they have different ideas of mathematical addition? Not at all. The classical view associates the addition of velocities (call this physical addition) with the simple addition of the associated real numbers, that is $(V_1, V_2) \rightarrow v_1 + v_2$. The champion of relativity adds velocities according to $(V_1, V_2) \rightarrow (v_1 + v_2)/(1 + v_1 v_2/c^2)$. (In our low-velocity example, this gives a value of 899.99. . . k.p.h. At much higher velocities the discrepancy between the two ways of calculating can become quite dramatic.) In each case '+' means the same thing. It is not the mathematical concept of addition, but rather the physical concept that has changed.

At first blush there are lots of bizarre things about the relativistic way of adding velocities, things which would puzzle us greatly if we, holding the classical view, encountered it for the first time in some exotic society. For instance, we can add velocities without a limit on how fast things can go. But on the relativistic view, no matter how many velocities we pile on other velocities, there is an upper limit of c. People selling lumber at different prices according to the shape of the pile – one of Wittgenstein's favourite examples – is certainly no more bizarre than that. And it would be sensible for us to attribute all bizarreness found to their views of nature, not to their views of mathematics.

An interesting sceptical problem can be raised here – though a different one from ours. The two ways of calculating the addition of velocities agree at low speeds (at least if relativistic calculators only bother with a few significant decimal places). We could not tell from actual calculations at low velocities which method of calculating is being used. This looks similar to our earlier example where teacher and student seem initially to be doing the same thing but then diverge as they further develop the sequence. However, the two problems are quite different. Our main problem is one concerning the understanding of mathematical rules. Scepticism in the other case is about physics – which physical theory do they hold?

Naturalism

Barry Stroud (1965) agrees with Dummett that radical conventionalism is a hopeless view of mathematics. But he disagrees with attributing radical conventionalism to Wittgenstein. It's true that we cannot imagine forms of inference radically different from our own, so it's not surprising that Wittgenstein's examples are 'thin and unconvincing' as Dummett puts it. Moreover, this is a good reason for thinking that radical conventionalism is false, otherwise we would have no trouble spelling out different possibilities. Nevertheless, we can imagine *that* such people exist, says Stroud, and this is enough to make plausible Wittgenstein's point that different ways of doing things are indeed possible.

I already mentioned above that these bizarre examples may be based on rival physical theories rather than different ways of doing mathematics. This could be developed in a way that undermines Stroud. However, I will ignore this point here, and instead focus on another.

Stroud remarks that 'it is a "fact of our natural history" in Wittgenstein's sense that we agree in finding certain steps in following a rule "doing the same". In some cases we all naturally go on in the same way from steps which have already been taken. This is what makes it possible for us to follow any rules at all' (1965: 494).

Is there any explanation for this? Apparently not. As Wittgenstein put it: 'The danger here, I believe, is one of giving a justification of our procedure when there is no such thing as a justification and we ought simply to have said: *that's how we do it*' (*RFM*, III-74). 'What has to be accepted, the given, is – so one could say – *forms of life*' (*PI*: 226).

This may seem a shaky business, but Stroud thinks it is quite stable and reliable.

> Logical necessity is not like rails that stretch to infinity and compel us always to go in one and only one way; but neither is it the case that we are not compelled at all. Rather, there are the rails we have already travelled, and we can extend them beyond the present point only by depending on those that already exist. In order for the rails to be navigable they must be extended in smooth and natural ways; how they are to be continued is to that extent determined by the route of those rails which are already there.
>
> (Stroud 1965: 496)

Those who reject Platonism, but still want a kind of objectivity about mathematics will find this picture attractive – but only initially. On reflection, I don't see how it could be either the truth about mathematics or even the right way to understand Wittgenstein. The idea of a 'smooth and natural way' to extend some mathematical operation seems contrary to everything Wittgenstein holds. If there were such a thing as a 'smooth and natural way', it would be just as mysterious as any Platonic entity. So rather than say of a sequence that it exists only as far as we have calculated, but that there is a 'smooth and natural' continuation, we might as well simply say that the series already exists infinitely far.

Consider Stroud's analogy of the railway for a moment. Why is one family of extensions of the existing rails 'smooth and natural' while another extension is not? The answer has to do with laws of nature. Inertia, for instance, is such that if a bend in the tracks is too sharp, then the train will fall off. These laws, which are utterly objective, non-conventional, facts of nature, are the constraints that make some extensions smooth and natural. For Stroud's analogy to work, there would have to be a mathematical counterpart to these laws of nature – a Platonic assumption, if ever there was one.

As for Wittgenstein, his sceptic denies that there is anything about the series so far, or anything the questioner has said, done or intended that determines the next element of the sequence. If there were a 'smooth and natural way' then this would be a fact which determines (if only with some probability) what the next element is. This completely flies in the face of Wittgenstein's sceptical point.

The Sceptical Solution

Kripke remarks that 'Wittgenstein has invented a new form of scepticism.' And he regards it as 'the most radical and original sceptical problem that philosophy has seen to date' (1982: 60). Much of Kripke's discussion is modelled on Hume whose scepticism concerned causation and induction. Is there anything in past events (such as a causal nexus), which determines future events? And is there anything in our knowledge of the past that allows us to make correct inferences to the future? These are related problems. Hume's answer to both is No, and herein lies his scepticism. But Hume also offered a sort of solution which he called 'sceptical': Causation is just regularity – nothing more; our inferences about the future are based on custom and habit – nothing more. Kripke notes the analogy between Wittgenstein and Hume and likewise calls Wittgenstein's approach a 'sceptical solution', since it allows us to carry on with our normal practices even though it concedes the seriousness of the problem. (This is in contrast to a 'straight solution' which tries to refute or in some way reject the sceptical problem.) In summary, Kripke's reconstruction of Wittgenstein runs as follows:

- There is no fact of the matter about whether the teacher/questioner meant one sequence rather than another.
- The only way to solve the problem is to abandon the view (found among realists such as Frege and in Wittgenstein's earlier *Tractatus*), that meaningful sentences purport to correspond to facts. Language does not work by being representational.
- However, this is not to abandon language. There are many useful things that various 'language games' can do.
- The conditions for the use of any language game involve reference to a community, all of whom play the same game.
- Ultimately, we act without hesitation, without justification; but this is not to act wrongly. In following a rule we simply do what we think is

right. There is nothing deeper. But it is not enough to do merely what we think is right. 'To think one is obeying a rule is not to obey a rule. Hence it is not possible to obey a rule "privately"; otherwise thinking one was obeying a rule would be the same thing as obeying it' (*PI* §202).

- We are trained by others, and we are judged to be following a rule correctly when we give the same answers as other members of the community. There is a strong tendency for all members of the community to give the same answers – a shared 'form of life'. This is just a brute fact, like Hume's regularities in nature. There is no explanation for this, no hidden cause: 'What has to be accepted, the given, is – so one could say – *forms of life*' (*PI* §226).

- Our answers are (in some loose sense) in accord with the rule; but they are *not caused* by it. We do not grasp a rule which then determines our behaviour; we do not agree in expanding a sequence because we share a common conception of some mathematical function. Rather, we say that we share a common conception because we agree in our answers, we agree in how we go on expanding the sequence.

As mentioned at the outset, there are three concerns: What is Wittgenstein's view? What should he have said, given his other views? And, is either of these true? Kripke's analysis is one of the interesting reconstructions of Wittgenstein's actual view. It has been repeatedly criticized, often with force.[3] But it has not been replaced with a more plausible alternative account. Let's accept it (at least tentatively) as the right way to understand Wittgenstein's actual view, so we can now get on to the other two questions.

Modus Ponens or *Modus Tollens*?

Philosophers are fond of paraphrasing an old saw: One person's *modus ponens* is another's *modus tollens*. Wittgenstein started from anti-Platonistic assumptions, developed a sceptical position, then inferred (*modus ponens*) the remarkable conclusion that there is no explanation for why we tend to agree in our answers; it's just a brute fact that we do.

Others – and I include myself – will find this conclusion wholly absurd and think it ridiculous to say that it is only an amazing coincidence that we all agree in developing a sequence in the same way. They will infer (*modus tollens*) that something is wrong with the premises of any argument that leads to this absurd outcome, and so conclude that there are, after all, independently existing mathematical entities and we can grasp them, just as the Platonists have been claiming all along.

When Hume claimed that we cannot make rational inductive inferences about the future, he put his finger on an interesting problem. But no sensible person should accept his sceptical conclusion. At most, Hume's discovery is an inducement to re-examine the premises of his argument. The proper response to Wittgenstein's argument is similarly clear. However, Wittgenstein might not be claiming there is nothing more than a remarkable coincidence in our answers; he may really be up to something else. It's time to speculate.

What is a Rule?

So far we have not said what a rule is; we have relied on an intuitive understanding. Let's spell things out in a bit of detail. The following seem to be necessary (though probably not sufficient) conditions for being a rule:

(1) A rule must apply to an indefinite number of situations. An *order* should be distinguished from a *rule*, though both can be 'followed'. An order is typically a one-time only thing: 'Close the window.' A rule covers indefinitely many instances: 'When it's cold out and the room is draughty, close the window.'

(2) A rule must be something that finite creatures like us can grasp in spite of the fact that they apply to infinitely many different cases.

(3) A rule is capable of guiding our actions; in some sense a rule *causes* our behaviour. Having grasped the rule and wanting to comply with it, we are typically able to do so. (This is a fallible process and quite compatible with making the occasional mistake in applying the rule.)

(4) It is possible to act *in accord* with a rule without following it. I might, for example, make some random noises which turn out to be a grammatically correct sentence of some language I've never heard of. In doing so I act in accord with the grammatical rules of that language, but I do not follow them. Again to adopt the causal idiom, to follow a rule is to have our action caused by that rule, and not to be merely in accidental accord.

When we reflect on these conditions, it becomes pretty clear that the 'sceptical solution' is rather unsatisfactory. In effect it denies that there are any rules. Wittgenstein's solution is not a surprising claim about the detailed nature of rules, but rather a highly implausible assertion that rules don't even exist. (The parallel with Hume is still intact, since Hume, in effect, claimed that there are, after all, no causes and no laws of nature. And that's why there are no rational inferences about the future.) Clearly, a different account is needed, one which takes rules to exist.

There is a fairly elegant way to account for rules, as characterized above, using Frege's distinction between *sense* and *reference*, identifying the grasping of a rule with the grasping of a Fregean sense. A term (a word, a sequence of symbols) has a *reference*, the object that it names. The reference of 'the famous Austrian philosopher who worked in Cambridge and wrote the *Tractatus*' Wittgenstein. The reference of 'the sequence n^2' is a set of ordered pairs with infinitely many members. These terms also have a *sense* 'wherein the mode of presentation is contained' (Frege 1892: 57). The terms 'the morning star' and 'the evening star' have distinct senses even though they have the same reference, namely Venus. The terms 'leprechaun' and 'tooth fairy' have the same reference, namely nothing at all, but clearly have a different sense. For Frege, senses are perfectly objective, something that anyone can in principle grasp. They exist quite independently of us; a type of Platonic entity. (Besides Frege, others have attempted a roughly similar distinction using the terminology: *intension/extension* or *connotation/denotation*. The word 'meaning' is often used ambiguously among these.)

Now consider a rule, mathematical or other. We may take it to have a reference, say, an infinite set of ordered pairs. We may also take it to have a sense. The claim now should be readily clear: Grasping a rule is grasping its sense. True, a rule covers infinitely many cases, but it is the reference of the rule that has infinitely many instances. That our limited minds cannot behold. But we can grasp the sense of the rule, and that is all we need to grasp in order to understand it and to apply it to new cases.

Is this a Platonistic account of rules and rule following? Certainly. Is it opposed to everything Wittgenstein stands for? Not as much as one might think. In fact, there is a way of taking this that is surprisingly Wittgensteinian in spirit.

Grasping a Sense

Postulating a sense of a rule as well as its reference or extension solves a problem, but raises others. One problem that I won't take up is how we might get in touch with this wierd abstract entity. That's a general problem for any sort of Platonism, but it was dealt with earlier (in Chapter 2). For our purposes here, I will simply assume that we can somehow or other grasp the sense of a rule.

The problem that I want to focus on has to do with the complexity of a sense. Discussions of these issues tend to assume either that we can or that we cannot grasp a sense, as a matter of principle. It's all or nothing. But, in reality, things are likely to be much more complicated. We seem to have no trouble grasping the sense of a simple rule, say the rule that tells us to continue a sequence by squaring the successive integers. (Several years of study with gifted tutors seem necessary to work up scepticism for this sort of example.) But what about much more complicated rules? There is really no upper bound on how complicated a rule can get. As rules become increasingly complex, their sense becomes increasingly difficult to grasp. At some point it will become humanly impossible to grasp the sense of a particular rule.

This is not a mere philosophical contrivance conjured up for the purpose at hand. The notion of 'complexity of a rule' is crucial to some mathematical concepts. Consider the idea of a random sequence of zeros and ones, say, something like this:

0, 0, 1, 0, 1, 1, 0, 1, 1, 1, 0, 1, 0, . . .

We think such a random sequence is lawless, not governed by a rule. Yet this is not so. For any such sequence there is a rule that covers it. Because of this, it has proven very difficult to give a characterization of randomness. One clever idea which is now quite popular is to characterize a random sequence in terms of the complexity of the rule which generates the sequence. Intuitively, the idea is this: a sequence is *random* when the rule which generates the sequence contains more bits of information than the sequence itself. For example, suppose we have a sequence of zeros and ones that has 1,000 terms. We could just write out the sequence. Or we could write out the rule that generates it. If the rule is as long or longer than the sequence itself, then the sequence is random. (The idea allows 'degrees of randomness' as well.)

If a 1,000-member sequence looks like this:

0,1,0,1,0,1, . . ., 0, 1

a simple, short rule will generate it.

    (1)   Write a 0
    (2)   Write a 1
    (3)   Repeat 1 and 2 500 times.
    (4)   Stop

But if the sequence looks like this:

0, 1, 1, 0, 1, 0, 0, 1, 0, 1, 1, 1, . . . 0, 1

the rule would be:

    (1)   Write a 0
    (2)   Write a 1
    (3)   Write a 1
    (4)   Write a 0

⋮

    (1000) Write a 1

    (1001) Stop.

The rule is longer than the sequence itself.

Normally we have a good grip on the sense of the terms we use, say, 'tree' or 'the sequence n^2'. It's the reference or extension of these terms that eludes us. We can exhibit a finite number of instances of n^2, but not all of them; and we can admire or chop down some trees, but not all past, present and future trees to which the term refers. In the case of random sequences it's the other way round; we have a better grip on the reference than on the sense of the rule. In the case of an infinite random sequence, we have a full grasp of neither – a partial grasp at best.

Let's agree then that rules are of varying complexity and that, consequently, the senses of these rules can be grasped with varying degrees of difficulty – some perhaps not at all.

Wittgenstein tried to make his case with examples like 'adding two'. His scepticism seems bogus and implausible. And it seemed so both initially and after detailed consideration of Wittgenstein's position. However, the plausibility of his case would be much greater if he had focused on complex examples. For instance, contrast expanding the sequence n^2 with expanding a sequence according to some rule based on, say, the computations involved in the proof of the four-colour theorem, with each successive element of the sequence requiring a much greater mastery of the four-colour proof. The rule for calculating successive terms of the sequence would become absurdly complex.

If we assume that rules can be indefinitely complex and only partially grasped, then the application of such rules to new cases will almost certainly lead to ambiguities. If the sense of a highly complex rule is only partially grasped, then at some point it will be certainly true that the holder of the rule had nothing in mind that determines 'the next member of the sequence'. This would hold *even though both the sense and the reference of the rule each have an independent existence.*

So, we can generate a Wittgensteinian problem about rule-following even in a highly Platonistic setting. But what about the Platonism itself? Just how un-Wittgenstein is it?

Platonism versus Realism

Platonism and realism are usually run together. This is perfectly natural, but the doctrines are actually distinct. Platonism asserts the existence of abstract entities; it says that numbers, functions, rules and so on are just as real as trees and electrons, though they are not physical entities located in space and time. Realism, on the other hand, is the doctrine that statements about numbers, rules, trees, and electrons are true (or false) independently of our beliefs, our evidence, our conceptual structure, our biology, our ways of testing.

In recent years, realism, as I've just characterized it, has come to be called metaphysical realism or external realism. It is commonly distinguished from, say, scientific realism or internal realism, which says that the theoretical entities of science (electrons, genes) have the same ontological status as observable entities (trees and cloud chambers). The observable/theoretical distinction might be epistemically important but, according to scientific realism, it is ontologically irrelevant – electrons are just as real as trees.

Verificationism is opposed to metaphysical realism; it need not be opposed to scientific realism, or to Platonism. One might say that any true statement is ultimately true in virtue of how we come to believe it. 'True' means confirmed under ideal epistemic circumstances or rationally believed at the end of inquiry. Thus the likes of Kant, Pierce and Putnam would have no trouble with believing in electrons. They are just as real as trees, provided that the electron theory is confirmed under ideal circumstances. But neither electrons nor trees are part of the 'noumena', neither are metaphysically real, neither are 'really' there completely independent from us, their knowers.

Exactly the same could be said of abstract entities. Numbers, functions and rules are just as real as trees and electrons, though not metaphysically real. To say they are real is just to say that under ideal conditions theories involving abstract entities are confirmed in the same way any other theory is confirmed. Thus Platonism is like scientific realism. Both can be taken in the internal realist way and be sharply distinguished from metaphysical realism.

Historically, I think Pierce held such a view. On the one hand, he identified truth with what is held at the end of inquiry – truth and evidence are linked. On the other, he was at pains to stress the existence of something he called 'thirdness', an abstract principle which has causal powers, but is certainly not in space and time (see, e.g., Pierce 1957). Perhaps Crispin Wright also holds something like this. He has clear verificationist sympathies (Wright 1993) yet, at the same time, he embraces Frege's Platonism in number theory (Wright 1983).

Whether there are historical precedents is of no matter. The point is that there is a distinction to be made between realism and Platonism. And this can be applied to Wittgenstein. It's true he often said that he rejected Platonism, but in the light of the distinction, it's actually metaphysical realism about abstract entities that is being rejected, not abstract entities themselves.

One could consistently adopt Platonism and reject realism, but I should stress that I do not advocate such a view myself, though the distinction is real. I remain convinced of the truth of both. I am only trying to foist this view on Wittgenstein.

Surveyability

Wittgenstein repeatedly stressed the importance of 'surveyability', a crucial notion for him. A proof is surveyable when we can grasp it, we can take it in as a whole. The notion is not easily defined, but it is readily understood in an intuitive way and we have no trouble applying it to various examples. The standard proofs of the Pythagorean theorem or the proof of the irrationality of $\sqrt{2}$ are surveyable, but the computer proof of the four-colour theorem is not.

This is exactly the notion we need for our account of a rule's sense. The sense of some rules is surveyable while others are too complex. Graspable and surveyable amount to the same thing. Here is a natural place for Wittgenstein's anti-realism to come into play. There are no rules with a transcendent sense; the only legitimate sense – Platonic entity though it is – is a surveyable sense.

In keeping with typical verificationist notions, by *surveyable* I mean something like *surveyable under ideal conditions*. Someone may have only a partial grasp of the sequence n^2, nevertheless, there is a definite infinite sequence of terms which are determined by this rule. But a sequence governed by a rule so complicated that no one could grasp it under any circumstances, however ideal, simply does not have its terms already determined and existing independently of us. In short, real = verifiable = surveyable. For example, there is a short (hence surveyable) rule for calculating the value of π to any decimal place. Thus the infinite expansion is perfectly real and already exists independently of our actually carrying out the calculation. By contrast, there is no rule for locating the twin primes – we simply have to work out the instances. Thus, by this criterion, there is no fact of the matter about twin primes, though there is about the expansion of π.

This quasi-Wittgensteinian account seems to me a great improvement over Wittgenstein's initial view. It has the virtues of allowing a kind of objectivity demanded by common sense in the expansion of a sequence, yet, at the same time, it recognizes real human cognitive limitations. It has the drawback of countenancing abstract entities, and for many commentators this will be enough to scuttle the view. Perhaps Wittgenstein himself would dismiss it for this reason as well, in spite of the distinction I drew between realism and Platonism. But there is no Wittgensteinian reason for doing so. Wittgenstein would certainly not allow philosophers *qua* philosophers to say that electrons or other unobservable entities of science do or do not exist in principle. So why should he, on philosophical grounds, dismiss abstract entities? Anti-realism is certainly part of his core theory. But it does not follow from this that the only things we meaningfully talk about must be material objects in space and time. There is room in his view for abstract entities, too.

There are passages in the *Remarks* that express a kind of wistful ambition: 'I should like to be able to describe how it comes about that mathematics appears to us now as the natural history of the domain of numbers, now again as a collection of rules' (*RFM*, IV-13). Perhaps this type of (anti-realist) Platonism does the trick.

The Sense of a Picture

How do we come to grasp the sense of a rule? Children somehow manage to grasp addition, multiplication and so on through examples: adding apples, making change, memorizing the multiplication table. These activities seem to trigger a grasp of the appropriate rule. One of the more interesting ways in which we often grasp a rule is by means of a picture. This is hardly ever taken up in discussions in the philosophy of mathematics, yet Wittgenstein dwells on it to a considerable extent, as I mentioned at the outset.

Our initial problem, recall, was to make sense of the flawed picture: How is it possible to grasp correctly what is going on when the pictorial representation is flawed?

Once again, here is the (incorrectly drawn) diagram (Figure 9.3) together with some of Wittgenstein's remarks.

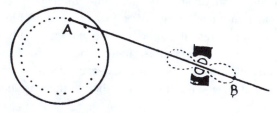

Figure. 9.3 Wittgenstein's mis-drawn diagram

> While the point *A* describes a circle, *B* describes a figure eight. Now we write this down as a proposition of kinematics . . . we can say that proof must shew the existence of an internal relation . . . The picture (proof-picture) is an instrument producing conviction.
>
> (*RFM*, VII-72)

When we consider examples like this, one thing seems clear: pictures, like rules, have a sense, or, at least, they enable us to grasp the sense of some associated proposition (in this case, the proposition that when *A* moves in a circle, *B* moves in a figure eight). But before we have the full solution we need one more important ingredient.

Recall a distinction made in Chapter 3. Those in aesthetics and perceptual psychology sometimes distinguish between two types of representation: a 'picture' and a 'symbol' (Arnheim 1969). A painting such as David's *Napoleon* is a representation, *a picture*, of Napoleon, but it is also a representation, *a symbol*, of more abstract things, such as: courage, leadership, glory, adventure. With this distinction in mind, it is easy to see that a particular representation might be a poor picture but a good symbol, or vice versa. Our particular drawing is thus a poor picture in that it is a mis-drawing of an actual mechanism. But it is a good symbol in that it represents the important abstract feature; it 'shews the existence of an internal relation', as Wittgenstein put it. In this latter respect it is quite successful, indeed.

And that, very simply, is the solution to the problem. The picture, flawed though it is, is simply an aid to the understanding; it triggers the grasping of the sense. Thus, as a picture it need not be accurate; it need only lead to the sense, the Platonic entity.

There is a bonus with this solution. Recall that I conceded above that some rules might be so complex that we could not fully grasp their sense. Perhaps we need not make this concession. If we allow ourselves the possibility of grasping senses via pictures as well as via the usual linguistic means, then it is far from clear how to draw the line between the surveyable and the non-surveyable. That is, we can grant that a rule does not have a determinate extension unless it has a (in principle) surveyable sense, but we need not grant that there are unsurveyable senses, since there is no known limit on the 'motley of techniques' that could in principle be available to us. Human ingenuity could perhaps always find a way of grasping even the most complicated rule's sense – any rule can be followed.

Further Reading

Wittgenstein's writings on mathematics can be found in several places, but *Remarks on the Foundation of Mathematics* is the most important. Wright's *Wittgenstein's Philosophy of Mathematics* discusses it in great detail. There is a huge literature on Wittgenstein, much of it on his mathematical views. A recent general anthology is Sluga and Stern, *The Cambridge Companion to Wittgenstein*. Kripke's *Wittgenstein on Rules and Private Language* is of central importance. An excellent biography is Monk, *Ludwig Wittgenstein: The Duties of Genius*.

CHAPTER 10
Computation, Proof and Conjecture

The Four Colour Theorem

How many colours does it take to do a map so that adjacent countries are different colours? The answer is so well known that hardly more than a sentence or two is needed to remind us of what it says. There are a few conditions: the map must be on a plane or sphere; the number of countries is finite; they must be connected (the USA, for example, isn't since Alaska and Hawaii are separate from the other states); and countries meeting only at a vertex may be the same colour. This has long been a popular problem, and those who tried their hand at it were usually convinced by experience that four colours suffice. In 1976 Appel and Haken proved what was widely suspected; their celebrated result is now known as the four colour theorem (4CT). Its fame rests partly on the fact that it solved an outstanding problem; but even more, its celebrity resides in the way the theorem was established – a computer played an essential role in the proof.[1]

One of the first and most influential commentaries was provided by Thomas Tymoczko (1979) who claimed that we here have a new way of doing mathematics. According to Tymoczko, computer proofs, such as that of the 4CT, are *not* a priori, are *not* certain, are *not* surveyable, and are *not* open to double-checking by other mathematicians. This makes computer proofs quite different from the traditional kind. They are, Tymoczko stresses, perfectly legitimate proofs, but their effect is to make mathematics more like the empirical and fallible natural sciences. Let's look briefly at each of his points.

Traditional proofs are a priori, says Tymoczko, in the sense that they do not rely on empirical evidence. But this is not so in the case of computer proofs. We need to know how the hardware works and that it's reliable; how the software works and that it's not buggy. This is empirical knowledge of the non-mathematical realm and it is unheard of in traditional mathematics. But as soon as we admit we need this, we must inevitably admit further that *certainty* has been abandoned. We may have a very high degree of confidence in the results, but resting as they do on our experience-based beliefs about the hardware and software, we must concede fallibility.

Tymoczko's claims about the non-surveyablity of the 4CT are perhaps his most controversial. *Surveyability* is an important, though fuzzy notion. (It was discussed in the last chapter.) We can survey a proof when it's short and easy to grasp. But when a proof is much longer and more complex, we lose our grip; it ceases to be surveyable. All traditional proofs are surveyable – perhaps not by everyone, but at least by those with the appropriate skill and training. However, in the case of the 4CT, the proof is so long and complex that no human can grasp it. It goes without saying, therefore, that no human could double-check the proof to see if it is mistake-free. Putting these points together, Tymoczko concludes that computer proofs are something new on the mathematical landscape. It's a new way of doing mathematics, a way that makes mathematics empirical, probabilistic, and generally more like the inductive natural sciences.

Fallibility

Of course, mistakes in calculation are commonplace. Some mathematicians even take pride in being poor at simple arithmetic, just as some writers brag about their inability to spell. But this sort of mistake is not what Tymoczko has in mind. Ptolemy was wrong, and so was Newton, but not because they made calculation errors. Their mistakes are the mistakes of empirical science – false hypotheses. The fallibility of computer proofs stems from this source. We hypothesize certain things about a computer and about the software that it's running. And this hypothesis may be wrong. Greater use of the machine in a variety of circumstances and greater testing of the software (or fragments of it) can lead to greater confidence, but it will never leave us with anything better than a well-confirmed scientific theory.

In the striking result of Lam *et al.* (1989) there is an explicitly probabilistic argument as part of the proof. They used several thousand hours of CRAY supercomputer time to prove the non-existence of finite projective planes of order ten. (It's not important to know what this is, but the seven-point geometry example cited in Chapter 5 is an example of a projective plane of order two.[2]) The computation required the examination of 10^{14} cases, which allowed plenty of scope for problems. In their paper, Lam *et al.* note the possibility of error. Interestingly, they note several types. One source of error stems from human mistakes in entering data. This is, perhaps, akin to regular calculation mistakes that are often made.

A second source stems from software problems. As we all know from regular commercial software, eliminating bugs is no easy matter. In writing a programme there are many levels of software to worry about: the operating system such as DOS or UNIX, the programming language such as PASCAL or FORTRAN, a compiler, and, of course, the particular software created to do the speciality computation. There are numerous opportunities for problems to enter. And with so few users, speciality software, in contrast to popular commercial programmes, has fewer opportunities to have its problems revealed.[3]

Third, one of the most interesting sources of error stems from hardware problems. In normal PC use hardware problems are glaring – a power failure, for instance. But some mistakes are random and possibly undetectable. In the case of CRAYs, mistakes are made on average about once per thousand hours. These are quite random, perhaps due to cosmic rays.[4] What Lam *et al.* do is (in brief) note the frequency of this type of error, the total time elapsed in a computation, and the number of cases that had to be examined; they then calculate the chance that an error was made when the computer was examining a case that was actually an instance of a projective plane of order 10. They conclude that this probability is very small, and when some special considerations concerning the nature of the search are considered, the probability of a misleading error is 'infinitesimal' (Lam *et al.* 1989: 1122).

In describing what he takes to be the philosophical upshot of this work, Lam at one point says he is tempted to 'avoid using the word "proof" and prefer[s] to use the phrase "computed result" instead' (1990: 8). In the end he draws the same conclusion as Tymoczko: 'As physicists have learned to live with uncertainty, so we [mathematicians] should learn to live with an "uncertain" proof' (*ibid.*: 12).

The problem – if it is a problem – has been recently exacerbated by DNA computing. Typical computing is based on the silicon chip. But Leonard Adleman (1994) managed to solve a problem in graph theory by manipulating strands of DNA.[5] The required background assumptions are extensive, in the form of a great deal of biology and biochemistry. Probabilistic assumptions are entering at many levels.

What are we to make of this? For one thing, Tymoczko's conclusion seems perfectly correct. That is, empirical, fallible and probabilistic elements are part of mathematics. But it is not quite so new as Tymoczko claims. It's hard to see his point except as being on a continuum. We make simple mistakes of adding or multiplying. In longer inferences, we make logical mistakes. In a great many proofs that count as traditional mathematics, mistakes have been made – ironically, in the four colour case itself. In the nineteenth century Arthur Kempe 'proved' the result, and more than a decade passed before the subtle flaw was discovered. Most recently, Andrew Wiles's first go at Fermat's Last Theorem was not correct. Now, his revised proof is thought to be right; but is anyone willing to bet heavily on it?

Surveyability

Much of the probabilistic nature of the 4CT comes, according to Tymoczko, from the fact that it is not surveyable. A number of questions arise? Is it true that the 4CT is not surveyable? Are all regular proofs that are typically found in the journals surveyable? What is the distinction between surveyable and non-surveyable proofs, and it is epistemically important? As I mentioned above, the notion of being surveyable is vague. Here's Tymoczko's gloss:

> A proof is a construction that can be looked over, reviewed, verified by a rational agent. We often say that a proof must be perspicuous, or capable of being checked by hand. It is an exhibition, a derivation of the conclusion, and it needs nothing outside itself to be convincing. The Mathematician *surveys* the proof in its entirety and thereby comes to *know* the conclusion.
>
> (Tymoczko 1979: 59)

Must a proof be surveyable (in Tymoczko's sense)? What do we want from a proof? It cannot merely be a relation that exists between first principles and the theorem. It is first and foremost a piece of evidence that the theorem is indeed a theorem. Evidence is *evidence for us humans*. 'Us' might be restricted to trained mathematicians. Proofs must at least be convincing to them. Persuading an omniscient being isn't good enough. On the other hand, proofs don't have to convince my pet dog. If proofs are logical relations between premises and conclusions, then it must be a logical relation that we humans can grasp. If proofs are pictures, then the pictures in question must, for example, be visible to humans. All realists are happy to say that it is perfectly possible for a proposition p to be true even though no one knows or even could know that it is true. But evidence is not like this: evidence which is incomprehensible to us is simply not evidence.

Surveyability is much like observability. In the natural sciences there is a commonly drawn, but rough, distinction between what is observable, i.e. can be seen with the unaided eye (trees, streaks in a cloud chamber) and what is theoretical, i.e. cannot be directly experienced (atoms, genes). It is, perhaps, natural to extend 'seeing' to include atoms and genes when they are 'seen' with a powerful microscope. However, some theoretical entities (neutrinos, quarks) cannot be seen even in this extended sense. We detect them and come to know their properties only indirectly by means of other (already accepted) theories about how they interact with other observable entities. If the analogy with surveyability holds, then we would have, first, simple proofs that are surveyable, second, more complicated proofs that are surveyable in an extended sense, and third, proofs which simply aren't surveyable, and hence arguably aren't proofs at all.

The first kind needs no comment. The second is well illustrated by longer proofs where we can follow any given step, even though we have trouble grasping the whole. It would seem that the 4CT falls into this second category.

Long and complex computer proofs are not the only source of difficulty. One of the most interesting achievements in recent years is the classification of all simple finite groups.[6] The proof was carried out over many years, by a very large number of mathematicians. The number of journal pages is estimated at about 15,000. It became clear to Daniel Gorenstein (1979) that so much work had been done on simple finite groups that quite possibly a complete classification had been, or was very nearly, achieved. This is now a completely accepted result (see Gorenstein *et al.* 1994). Clearly, the result is not surveyable in the first sense, but would seem to be in the second.

Tymoczko mentions Lakatos with approval in connection with fallible mathematics. But there is an important distinction. The key to fallibility, according to Lakatos, stems from *conceptual change*. This is different from any type of 'mistake' discussed by Tymoczko. The concept of polyhedron changes with mathematical theorizing – earlier conceptions are tossed out. This is neither a calculation error nor a logical mistake, nor does the problem stem from an inability to survey the proof. Lakatos's is one more source of fallibility, but it is different from Tymoczko's.

Inductive Mathematics

Computer proofs may not lead to a revised view of mathematical evidence, just to a change in how we acquire some of that evidence. But there are other ways in which computers have come to play a major role in mathematical practice. In the next chapter I'll discuss some of the debates surrounding these issues. In the balance of this chapter, I'll take up the issue of computer-generated data and how we might consider it.

Mathematical rationality is based on much more than proof – whatever we think proof is. Mathematicians wonder about which problems to work on (or to give their students), and what techniques are most likely to succeed. They sit on grant-giving committees that evaluate the plausibility of various proposals, and they fund those thought sufficiently promising. As a body of *accomplishments*, mathematics may rest exclusively on proof (though this is highly questionable), but as an *activity*, mathematics depends heavily on hunch, plausibility and conjecture. We need only observe how mathematicians react to unsolicited manuscripts offering a 'proof' of a famous problem. Confident of them being flawed, usually they are put in the garbage. One of the more creative responses we sometimes hear of is to ask the author to put up a considerable sum of money on the condition that if the reader finds a flaw in the proof, she keeps the money; if she doesn't, she returns the money and publicly sings the author's praises. Is this arrogant dogmatism? I doubt it. Life is short while these 'proofs' are invariably long. 'Had we world enough and time . . .' it might be a different story, but in the real world of mathematics, inductive evidence can and must play an important role.

First of all, a distinction: an *open problem* is a proposition that so far has no proof. (Let us not worry about the nature of proofs for now; for the sake of the argument we can adopt the traditional conception.) Of course, we can make a *guess*, but a *rational conjecture* (or simply, a *conjecture*) is a statement of the open problem for which there is convincing evidence, convincing in the sense that a reasonable person would be inclined to believe a rational conjecture in the mathematical realm just as she would believe a similarly well-established theory in physics.

The distinction between rational conjecture and a mere guess is somewhat fuzzy, I admit, and the terminology is also a slight departure from normal usage (though Shanks (1993) notes a similar difference). So let me explain a bit further.

Normally, anybody can conjecture anything, in the sense of guessing – evidence has nothing to do with it. (Popper's whole philosophy of science is built on this.) But I want to mark an important distinction – vague though it is – and to tie rational conjecture to available, though partial, evidence in some objective way. It may be helpful to posit an ideal mathematician who, having all the available evidence (but not a proof), is rationally inclined to believe p, and thus conjectures it. If there is no evidence at all for or against p, then it's an open problem. Obviously, there is no natural boundary in the series: zero evidence, very weak evidence, mild evidence, strong evidence, and so on. And, of course, the availability of new evidence may push a rational conjecture p into an open problem or even into the rational conjecture $\sim p$.

So, the problem is now: What are the characteristics of mathematical evidence (other than proof)? What role does the computation of instances play? What other types of evidence besides computation could there be? What makes a mathematical conjecture a good one? What sorts of grounds could there be for accepting it?

Most topics in the philosophy of mathematics have a well-developed literature that serves as a point of departure for future reflection. But not here. Writings on this subject are sparse and quite underdeveloped. Consequently, I think it's a good idea to look at lots of examples. (Of course, it's always good to look at lots of examples, but here even more so than normally.)

Perfect Numbers

A *perfect number* is one which is equal to the sum of its positive divisors (other than itself). Thus, 6 is a perfect number since it is divisible without remainder by 1, by 2, and by 3, and $1 + 2 + 3 = 6$. The second perfect number is $28 = 1 + 2 + 4 + 7 + 14$. The Greeks knew the first 4; today about 45 are known.[7] Much mysticism has been associated with them; Augustine, for instance, thought that God took six days to make the world, as this signified the perfection of creation. Here are the first nine perfect numbers:

6

28

496

8,128

33,550,336

8,589,869,056

137,438,691,328

2,305,843,008,139,952,128

2,658,455,991,569,831,744,654,692,615,953,842,176

I won't list more since the size grows so quickly; the largest so far discovered has millions of digits.

Knowing only what the ancient Greeks knew (i.e. the first four perfect numbers on the list), one might conjecture (as was done in the middle ages) that the nth perfect number is n digits long, and that the list alternates between numbers ending in 6 and 8. But these conjectures were dashed by later discoveries. From a glance at the list, two other questions naturally arise: Are there *infinitely* many perfect numbers? Is there an *odd* perfect number? I'll focus on these questions, as they nicely illustrate the distinction between open problem and rational conjecture.

First, let's ask about the existence of odd perfect numbers. Are there any? (See Wagon (1985), Shanks (1993).) All known perfect numbers are even. But if there are infinitely many, this is a pretty meagre sample. Descartes thought there should be an odd perfect number, but he searched in vain. It is now known that if an odd perfect number exists, it must have more than 10^{50} digits and at least eight prime factors, one of which is greater than 300,000. So it's not too surprising that no odd perfect number has been found so far. Does the fact that there can't be any 'small' odd perfect numbers lend support to a conjecture that there are none at all? This, too, is doubtful. Perfect numbers tend to be 'large' anyway; from the tenth on they are all greater than 10^{50} (which is why I cut off my list at nine entries). In short, there is really no evidence at all to say that there is or is not an odd perfect number. There are no grounds for a rational conjecture here; this is an open problem.

Let's now turn to the question of how many perfect numbers there are. To do so, we must first look at *Mersenne numbers* which have the form $2^p - 1$ (where p is a prime). When $2^p - 1$ is itself a prime, it is called a *Mersenne prime*. The first few examples of Mersenne numbers are also Mersenne primes: $3 = 2^2 - 1$, $7 = 2^3 - 1$, $31 = 2^5 - 1$, $127 = 2^7 - 1$. It looks like they are growing in a regular and predictable way. But the next Mersenne number, $2047 = 2^{11} - 1$, is not a prime. Nevertheless, the next Mersenne number, $8191 = 2^{13} - 1$, is a prime, hence, a Mersenne prime.

Notice something remarkable: When we multiply a Mersenne prime by the next lower power of 2, we get a perfect number:

$$2^1(2^2-1) = 6$$
$$2^2(2^3-1) = 28$$
$$2^4(2^5-1) = 496$$
$$\vdots$$

In fact, it is generally true (a theorem) that if $2^p - 1$ is a Mersenne prime then $2^{p-1}(2^p - 1)$ is a perfect number. What's more, every even perfect number has this form. Thus, the existence of perfect numbers is tied to the existence of Mersenne primes. And, as you would expect, 45 are presently known to exist.[8] This means that we can focus our attention here when searching for evidence about the infinity of perfect numbers. And it is rewarding to do so, since looking at the list of perfect numbers we see little or no pattern, but the regularities in the Mersenne primes are striking. For example, look at the graph (Figure 10.1) of the base-two logarithms of the values of p which lead to Mersenne primes.[9]

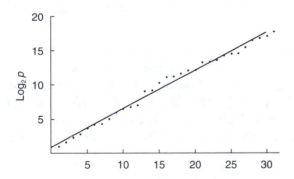

Figure 10.1 The distribution of Mersenne primes

The pattern is almost linear. It can be used to predict roughly where the next Mersenne prime will be located. It also suggests the existence of infinitely many such numbers. This, in turn, implies the existence of infinitely many perfect numbers.

Of course, this may just be a coincidence; at higher values of p the pattern may change abruptly. But it is the kind of consideration that mathematicians find plausible – and rightly so. Thus, the existence of infinitely many perfect numbers counts as a *rational conjecture*. Our belief is supported by some reasonably solid evidence. We do not have a proof, but our belief is rational, none the less.

Computation

Computation plays a huge role in the search for Mersenne primes. Computers are set to factor very large numbers, to see if they are primes or composites. Many took several hours of CRAY time to show that they are prime. Computation of this sort, however, is usually thought to be part of a normal proof. More interesting is computation which does not result in a proof. Goldbach's conjecture (and yes, it is a rational conjecture) says that every even number greater than 2 is equal to the sum of two primes: $4 = 2 + 2, 6 = 3 + 3, 8 = 5 + 3, \ldots, 1328 = 941 + 387, \ldots$. This has been checked well into the several millions, and holds invariably. It is widely believed. The huge number of examples seems to convince people.

But a huge number of instances is neither necessary nor sufficient, as a famous example of Littlewood's shows. Let $\pi(x)$ denote the number of primes equal to or less than x, e.g. $\pi(10) = 4$ since there are four primes, 2, 3, 5 and 7, equal to or less than 10. Then the integral

$$\int_0^x \frac{1}{\log t} dt$$

approximates $\pi(x)$. Littlewood proved that

$$\int_0^x \frac{1}{\log t} dt - \pi(x)$$

changes sign infinitely many times (thus, as an approximation, the integral alternates between overshooting and undershooting infinitely often). Yet when checked numerically, the integral is greater than $\pi(x)$ for every $x \leq 10^{12}$. The available *numerical* evidence was highly misleading. But we don't need fancy examples to begin to really appreciate the problem. Here's a strikingly simple one. Consider the following polynomial:

$$f(n) = n^7 - 28n^6 + 322n^5 - 1960n^4 + 6769n^3 - 13132n^2 + 13069n - 5040$$

When we test it, we see a definite pattern: $f(1) = 1, f(2) = 2, f(3) = 3$, and so on. It certainly looks as if $f(n) = n$, for every n. Yet the pattern breaks down at eight where $f(8) = 5048$, and never again does $f(n) = n$. The counter-example here could be found quite easily; we needed to examine only the first eight numbers. But it is interesting to see how the polynomial was constructed.[10] It was devised by simply multiplying out and collecting the terms in

$$f(n) = n + [(n-1)(n-2)(n-3)(n-4)(n-5)(n-6)(n-7)]$$

Each of the values of n up to 7 will, in turn, make one of the terms $(n - m)$ equal to zero, and so the result is due exclusively to the isolated term n. But when we exceed 7, this no longer happens. What is also clear is that 7 is quite arbitrary; there are indefinitely many different (though similarly devised) polynomials in which $f(n) = n$ as far as some arbitrary m and does not equal n thereafter. Thus, no matter how far we search, no matter how many millennia a CRAY works on the problem, there will be a polynomial that satisfies $f(n) = n$ up to the point we have examined, but diverges thereafter. Numerical evidence, then, is really no help at all in cases like this.

Those with a taste for Goodman's 'Grue' should find this sort of example intriguing. Let me explain. There is a temptation to think that inductive (or ampliative) inference is the same in mathematics as it is in science. But this is highly doubtful. Consider a function defined on the natural numbers. We may study this function for a very large number of arguments and claim to see a pattern in its values. However, for any f there is an f' which has exactly the same values for the examined input, yet diverges wildly for all input so far not considered. On the basis of the limited input and output examined, there is no way to say that the function in question is f rather than f'. There is no analogue of this in the physical realm. There does not exist, for example, a second species which looks black just like the raven in all cases which have been seen, while unseen ones are orange. (Of course, one could be sceptical about this, too.) It's a logical possibility, of course, but nature does not realize all logical possibilities – the Platonic realm does; it's as full as it can be. The assumed 'uniformity of nature' is often said to be a necessary assumption in inductive inference. This may be so, but I suspect that *the sparsity of nature* plays just as crucial a role.

Overlapping species may not exist the way infinitely many overlapping functions do, but there is a problem of induction raised by Goodman (1965) that has some surprising affinities – the infamous problem of *grue*. Emeralds are green; close observation over a very long time tells us this. At least, this is what we've inferred. But they might actually be grue instead of green. The colour *grue* is defined as *looks green up to some time t, then looks blue thereafter*. (Time t could be any future time, say, 1 January 2001, which some who can count claim is the first day of the new millennium.) All the empirical evidence we have so far accumulated about emeralds is compatible with them being either green or grue. And given this, we can't make rational predictions about the future appearance of emeralds, for if they are indeed grue then, after time t, they will look blue rather than green. (This, by the way, is perfectly compatible with the uniformity of nature; emeralds would be constantly and invariably grue.)

The problem of grue seems highly artificial. Perhaps it's of philosophical interest, but hardly a practical problem for working scientists. Surprisingly, it may turn out to have a genuinely interesting application in inductive mathematical inference.

Is π Normal?

Perhaps the most important and intriguing number in all of mathematics is π, usually defined as the ratio of the circumference of a circle to its diameter. Among its important properties is that it is irrational and even transcendental (i.e. unlike √2, it is not the solution of an algebraic equation). What is not known about π is this: is it normal? The intuitive idea behind normality is randomness; do the numbers in the expansion of π occur in a random way? But this is only a rough characterization; normal and random are different concepts. A real number is *normal* (in base *b*) if and only if in its representation all digits occur (in the limit) equally often, and strings of length *n* (for every *n*) occur equally often. Thus, if π is normal, then every digit must occur infinitely many times, and if the three-string 456 occurs *n* times (or infinitely many times), then every other three-string must occur exactly as often.

The first 50 digits of π suggest normality, since there seems to be no pattern at all:

3.14159265358979323846264338327950288419716939937511 . . .

But patterns can be more than a little deceptive here. The number 0.12345678910111213 . . . is normal (in base 10) in spite of exhibiting a very clear and predictable pattern. Nevertheless, seeing the expansion of π at much greater length would obviously be a help.

Computing π has long been a problem. As late as 1706 its expansion was known only to the first 100 digits (using the formula π = 16 arctan 1/5 − 4 arctan 1/239). And the ENIAC computer of 1949 still managed only 2037 digits after 70 hours. With faster hardware and a better formula[11] we can now do as well in the twinkle of an eye. Current expansions are now well into the billions. Most real numbers are normal (in the sense that the set of non-normal numbers is of measure zero), yet not a single example is known. (Even the example cited above is not known to be normal in any other base than 10.) So, what is the evidence that π is normal?

A statistical analysis on the first 10 million digits tells us something interesting. The frequencies for each of the ten digits are: 999,440 zeros; 999,333 ones; 1,000,306 twos; 999,964 threes; 1,001,093 fours; 1,000,466 fives; 999,337 sixes; 1,000,207 sevens; 999,814 eights; and 1,000,040 nines. All are very close to the expected one million. Perhaps even more impressive is the rate at which each digit is closing in on being 10% of the total. Consider the number 7, for example: It's 0% of the first 10 digits, 8% of the first 100, 9.5% of the first 1000, 9.7% of the first 10,000, 10% of the first 100,000, 9.98% of the first 1,000,000, and 10.002% of the first 10,000,000 digits. The speed of convergence to 10% is also at the rate predicted by probability theory (i.e. a rate proportional to $1/\sqrt{n}$).

These kinds of consideration lead one reasonably to expect that π is normal. Computation plays a crucial role; it provides essential data. But computation is not the end of the matter. It's the statistical analysis of the computed results that really provides the convincing evidence.[12]

Fermat's Last Theorem

Are there whole numbers, x, y, z, and $n \geq 3$, such that $x^n + y^n = z^n$? When $n = 2$, the equation is just the Pythagorean relation which is satisfied by lots of instances, such as $\{3,4,5\}$, i.e. $3^2 + 4^2 = 5^2$. But when $n \geq 3$, there are no known examples. In fact, Fermat thought he had actually proved that there were none. In his copy of a number theory book he wrote that he had a proof, but that the margin was too small for him to write it down. No one has been able to reconstruct what Fermat might have had in mind, and it is now generally agreed that he very likely had no proof at all. Nevertheless, Fermat's claim has been widely believed as a conjecture, if not a theorem (in spite of the fact that it goes by the name Fermat's Last Theorem, or FLT). Recently, events have taken a dramatic turn. In 1993 Andrew Wiles announced that he had proved the conjecture. Then it turned out that there was a significant gap in his proof. A year and a half later, the consensus was and remains that he (and a co-worker) have patched things up, and we do indeed have a proof of the theorem, hereafter likely to be known as the Fermat–Wiles theorem. It's worthy of inclusion in our discussion here since it was a famous conjecture that turns out to be right. (Below I'll look at one that didn't have a happy outcome.) Why did mathematicians believe FLT before having a proof?

Some special cases were proved early on. Fermat himself showed that FLT holds for $n = 4$; a century later Euler established that it held for $n = 3$; and about 75 years after that Legrendre and Dirichlet proved it for $n = 5$. Euler noted that his proof was so different from Fermat's that a method for attacking the general problem couldn't be extracted from them. The first significant breakthrough came in the mid-nineteenth century when Kummer (using his notion of *regular primes*) established FLT up to $n \leq 100$.

A different approach by Inkeri established that when n is a prime number, p, then, if there is a counter-example, $\{x,y,z\}$, to FLT, then x will have to be greater than $((2p^3 + p)/\log 3p)^p$. The latest computer-generated results on the size of n said that FLT holds at least as far as $n = 1,000,000$. This means that in any counter-example, x will be at least 17 million digits long.

As impressive as some of these considerations may be, they pale beside Gerd Faltings' results in 1983 (along with his proof of the Mordell conjecture) showing that there are *at most* finitely many solutions to the Fermat equation (i.e. at most finitely many counter-examples to FLT). Of course, that's compatible with there being billions of counter-examples; but his work is crucial for a different reason. Faltings' result shows that the case $n \geq 3$ is completely unlike the Pythagorean case where $n = 2$. The latter has infinitely many solutions. In this regard, there is no other number which is like 2.

Though Euler could not prove FLT, he did contribute greatly to it and to related problems. In fact, he conjectured a generalization of FLT that was also long thought to be true. The generalization said that if $n \geq 3$, then fewer than n nth powers cannot sum to an nth power. As a special case, this means that there are no solutions $\{w,x,y,z\}$ to the equation $w^4 + x^4 + y^4 = z^4$. The conjecture was well tested by examples, and for about two centuries was as widely believed as FLT. However, counter-examples have been found recently. For example, $2,682,440^4 + 15,365,639^4 + 18,796,760^4 = 20,615,673^4$.

The process of discovering this counter-example is of some interest. Brute-force computation would be hopeless with numbers this size. It came instead, via algebraic geometry, where the equation is associated with an elliptic curve. Certain properties of the curve suggest the existence of a counter-example, so a computer search is then unleashed, but only in the vicinity of numbers associated with specific points on the curve. This resulted in a manageable computer search, and the counter-examples were found. (Elliptic curves and their connection to modular forms play a central role in Wiles' proof of FLT.)

The Riemann Hypothesis

Now that the conjecture known as Fermat's Last Theorem is definitely a theorem (thanks to Wiles), the Riemann hypothesis must be the most famous conjecture in all of mathematics. It is surely the most important. In 1859 Riemann published his outline of a proof of the prime number theorem, which says that the number of primes less than or equal to the number x is roughly equal to $x/\log x$. There were lots of gaps (of which he was aware), and the final proof (by Hadamard and by de la Vallée-Poussin, independently) didn't come until the end of the century.

Riemann's method is based on the zeta function,

$$\zeta(s) = \sum_{k=1}^{\infty} \frac{1}{k^s}$$

The number s is a complex number, hence of the form (a,b), and often written as $a + ib$, where a is the 'real' part and b is the 'imaginary' part of the complex number. The *roots* of $\zeta(s)$ are the values of s for which $\zeta(s) = 0$. (Equivalently, the values of (a,b) for which $\zeta((a,b)) = 0$.)

The prime number theorem turns out to be equivalent to the assertion that there is no root (a,b) with $a \geq 1$. What is not known, but very widely believed, is the conjecture known as the Riemann hypothesis: all the complex roots of

$\zeta(s)$ have the form $(\frac{1}{2},b)$, that is, the real part of the complex number always has the value $\frac{1}{2}$. Among the relevant facts are these: There are infinitely many solutions of the form $(\frac{1}{2},b)$. (But, of course, there might also be some that aren't of this form.) The first 1,500,000,001 complex zeros of $\zeta(s)$ all have real part equal to $\frac{1}{2}$, and so do all the zeros after the first 10^{20} up to the first 10^{29} zeros. But there have also been some 'close calls' where ζ is almost 0 and the real part of s is not $\frac{1}{2}$. Some of these near misses have 'shaken' believers in the conjecture. I won't pursue the Riemann hypothesis further since examining the evidence requires too many advanced technicalities. I mention it here because it is the most famous and important conjecture in mathematics today. (For much more detail see Edwards (1974).)

Clusters of Conjectures

Often conjectures come in clusters. There may be evidence for each individually, but the fact that they are related to one another seems to lead to an additional kind of mutual support. The following cluster comes from number theory and among other things is related to odd perfect numbers, which were discussed above (see Ribenboim (1988) and Shanks (1993)).

(1) *There are infinitely many Mersenne primes.*
(2) *There are infinitely many Mersenne composites* (i.e. composites of the form $2^p - 1$, where p is a prime).

Conjectures (1) and (2) can't both be wrong; but they could both be right and indeed both are thought to be.

(3) *There are infinitely many primes $p = 4m + 3$ such that $q = 2p + 1$ is also a prime.*

There are lots of examples of this, including instances of p which are quite large, e.g. $p = 16188302111$. This conjecture implies a weaker one:

(4) *There are infinitely many primes p such that $q = 2p + 1$ is also a prime.*

This in turn seems closely related to the famous twin primes conjecture:

(5) *There are infinitely many numbers p such that p and $p + 2$ are both primes.*

There is a simple and quite convincing argument for (5). Primes seem to be distributed randomly, and there are infinitely many of them. So we should expect them to pop up over and over again spaced arbitrary distances apart, including being spaced two numbers apart. Thus, there are infinitely many twin primes. There is a stronger version of this conjecture which relates the distribution of twin primes in a way similar to the prime number theorem. It, in turn, is related in a rather complicated way to Goldbach's conjecture, which is, perhaps, the favourite example of most philosophers.

(6)   *Every even number (greater than 2) is the sum of two primes.*

Thus, $4 = 2 + 2$, $6 = 5 + 1$, $8 = 5 + 3$, . . ., $968 = 727 + 241$, . . . It's been tested by computer up to 10^8.

Polya and Putnam

Euler discovered that the sum of the series $1/n^2$ is equal to $\pi^2/6$. His argument was as questionable as it was ingenious. It's one of the few examples other than Goldbach's conjecture that has been discussed by philosophers. Hilary Putnam approvingly saw it as a wonderful example of 'quasi-empirical' methods at work.

> Euler, of course, was perfectly well aware that this was not a proof. But by the time one had calculated the sum of $1/n^2$ to thirty or so decimal places and it agreed with $\pi^2/6$, no mathematician doubted that the sum of $1/n^2$ was $\pi^2/6$, even though it was another twenty years before Euler had a proof. The similarity of this kind of argument to a hypothetico-deductive argument in empirical science should be apparent: intuitively plausible though not certain analogies lead to results which are then checked 'empirically.' Successful outcomes of these checks then reinforce one's confidence in the analogy in question.
>
> (Putnam 1975: 68)

Putnam follows Polya, both in the details of Euler's argument and in the morals to be drawn. Here's the chain of reasoning, according to Polya.

First, we note a fact (a theorem) about polynomials that Euler used.

$$a_n x^n + \ldots + a_2 x^2 + a_1 x + a_0 = (1 - x/r_1)(1 - x/r_2) \ \ldots \ (1 - x/r_n)$$

where r_i are the roots of the polynomial (i.e. the values of x that make the polynomial equal to zero). Next, Euler examined the Taylor series expansion of $\sin x$, setting this to zero.

$$\sin x = x - x^3/3! + x^5/5! - x^7/7! + \ldots = 0$$

The series has infinitely many terms, so if we think of it as analogous to a polynomial it should have infinitely many roots. And, of course, it's already known to have them, $0, \pi, -\pi, 2\pi, -2\pi, 3\pi, -3\pi, \ldots$ which tends to support the analogy. Euler tossed out the zero root, and divided the series by x (this would be the linear factor which corresponds to the zero root). Now we have the new series,

$$1 - x^2/3! + x^4/5! - x^6/7! + \ldots = 0$$

with roots: $\pi, -\pi, 2\pi, -2\pi, 3\pi, -3\pi, \ldots$. Consequently,

$$(\sin x)/x = 1 - x^2/3! + x^4/5! - x^6/7! + \ldots$$

which, by analogy with the normal polynomial case, equals

$$(1 - x/\pi)(1 - x/-\pi)(1 - x/2\pi)(1 - x/-2\pi)(1 - x/3\pi)(1 - x/-3\pi) \ldots$$

A bit of ordinary algebra yields

$$(\sin x)/x = (1 - x^2/\pi^2)(1 - x^2/4\pi^2)(1 - x^2/9\pi^2) \ldots$$

Another use of the analogy plus a bit more algebra gives us

$$1/3! = 1/6 = 1/\pi^2 + 1/2\pi^2 + 1/4\pi^2 + 1/9\pi^2 + \ldots$$

From this Euler derives the famous result

$$\pi^2/6 = 1 + 1/4 + 1/9 + 1/16 + \ldots$$

Polya then goes on to note that Euler calculated and found perfect agreement as far as he went. Moreover, he used the same style of reasoning to derive other results, such as Leibniz's series for $\pi/4$. These are the considerations that Putnam celebrates in saying 'intuitively plausible though not certain analogies lead to results which are then checked "empirically"'. Polya sums up his discussion, saying, 'Euler seems to think the same way as reasonable people, scientists or non-scientists, usually think. He seems to accept certain principles: *A conjecture becomes more plausible by the verification of any new consequence.* And: *A conjecture becomes more credible if an analogous conjecture becomes more credible*' (1954: 22).

Conjectures and Axioms

In broad outline, there are three different views one could adopt concerning axioms. First, that they are self-evident truths. This is a view commonly associated

with the history of Euclidean geometry. Second, that axioms are arbitrary stipulations. Conventionalism and formalism hold this view and say that we have
complete freedom in postulating whatever we want (provided we uphold consistency). Third, that axioms are fallible attempts to describe how things are. Gödel
and Russell held versions of this third view, a view I defended in Chapter 2. Like
the first option, it is a realist view, but claims no certainty. Axioms are conjectured
and, like scientific theories, tested by their consequences. Let's focus on this third
view and its relation to testing conjectures.

The axioms of set theory are conjectured and so is the Riemann hypothesis,
but there seems a world of difference between them. What's the difference? The
way evidence is marshalled for each is similar, so we cannot easily appeal to an
epistemic difference in their status. Instead, consider an analogous situation in
physics. Newton's laws of motion and the law of universal gravitation were
conjectured. So also was the existence of the planet Neptune. It was conjectured
by Adams and Leverier to explain the motion of Uranus. What's the difference
between these conjectures? I suspect that the answer is rather simple: The latter
conjecture presupposes the truth of the former. Adams and Leverier assumed
that Newton was right, and they then tried to fill in a detail or correct an anom-
aly. Their conjecture might even be seen as conditional: If Newton is right, then
there is a hitherto hidden planet located at such and such a place, etc. In other
words, the Adams and Leverier conjecture is *within* a framework, while
Newton's conjecture *is* the very framework itself.

The axioms for Euclidean geometry, or for set theory, are similarly conjectured frameworks, while the Riemann hypothesis is a conjecture within.
Goldbach's conjecture should be seen as being a conditional: Given the right-
ness of the Peano axioms, every even number (greater than two) is the sum of
two primes. Of course, real life is messy. Goldbach made his conjecture two
centuries before the Peano axioms were formulated. Nevertheless, it seems safe
to say he had similar arithmetic principles in mind, even if they were not made
explicit in his day.

But this can't be the whole story. Let's for a moment ask, How would the
Riemann hypothesis stand, if it were shown to be independent of the axioms of
set theory? Realists, of course, still would want to know whether it is true or
false (and much of the current evidence would likely remain relevant), but a significant number of mathematicians would lose interest in the problem altogether (or, perhaps more likely, develop a different kind of interest in it). What
such a reaction shows is that our interest in conjectures which are not axioms is
twofold: we want to know if they are *true*; but we also want to know if they are
derivable. This, I think, is the key difference between ordinary conjectures and
conjectured axioms.

Aside from proofs, the notion of mathematical inference is a largely unexplored field. It is certainly not in the same stage of development as, say, rational
inference and methodology in the natural sciences. There are some exceptions,

however. Penelope Maddy has thrown considerable light on the processes involved in coming to accept axioms in set theory (see Maddy 1990: especially ch. 4). This is similar to what I called framework conjectures. As for conjectures within a framework, Daniel Shanks (1993) is one of the very few to discuss what *generally* makes for good ones, though his attention is confined to number theory. I hope that raising these issues here will stimulate readers, following the lead of Polya, Putnam, Maddy and Shanks, to think seriously about the various forms of mathematical evidence.

Further Reading

Polya's various writings on heuristic reasoning are always interesting and full of insight. Start with *Mathematics and Plausible Reasoning* (2 vols.). Maddy's *Realism in Mathematics* has lots to say about conjectures in set theory. Shanks's *Solved and Unsolved Problems in Number Theory* is very good on that particular topic. *New Directions in the Philosophy of Mathematics* (ed. Tymoczko) contains several papers related to this topic.

CHAPTER 11
How to Refute the Continuum Hypothesis

New mathematical intuitions leading to a decision of such problems as Cantor's continuum hypothesis are perfectly possible.

Gödel

CH is *obviously* false.

Cohen

Cantor's continuum hypothesis (CH) is one of the great outstanding problems of modern mathematics. Hilbert made it number one on his famous list of problems in 1900. After decades of trying, it turned out to be a hopeless task. Gödel and Cohen showed it to be independent of the other axioms of set theory. And yet, the question of its truth remains open. It may have been settled in the negative by Chris Freiling, but his 'refutation' has gone largely unnoticed, perhaps because it was by means of a remarkable thought experiment, a method that is far removed from common approaches, but one that would get a sympathetic hearing from those who like picture proofs. By fleshing out some of the details, perhaps we can show it in a favourable light. This might in turn generate some serious interest in the result itself and in the unusual method used to achieve it. After a few more introductory remarks, I will explain the result in detail.

If Freiling's approach works, it will be a huge vindication of visual methods in mathematics. But even if it is a failure – and it may well be – it still sheds considerable light on CH and on the potential power of visual reasoning. His method of argument is alien even to those working in foundations. Looking for new axioms is a commonplace among foundational mathematicians, but the method of doing so is akin to Gödel's recommended consequentialism; that is, they posit a new axiom, then check its logical consequences. New axioms are supported by the fact that they are, as Gödel put it, 'abundant in their verifiable consequences, shedding so much light upon a whole field, and yielding such powerful methods for solving problems . . .' (1947/64, 477). This is a much more liberal attitude to the discovery of mathematical truths than that held by most working mathematicians, but it is still wedded to *derivations* as the one and only way to establish verifiable consequences. Liberal though this view may be, it is not so liberal as to include thought experiments involving probabilistic outcomes as a legitimate method of justifying mathematical propositions. This, I suspect, is why even those who work on the foundations of set theory ignored Freiling's proof. Admittedly, these are rather speculative explanations for the rejection of Freiling's work.

David Mumford is a notable exception. Mumford is one of the great mathematicians of recent times. His early work was in algebraic geometry, for which he won a Fields Medal in 1974. More recently, he has become interested in stochastic mathematics. It was this that led him to Freiling's work, which he claims is as important as Gödel's. Mumford is quite ambitious in his plans for mathematics, wanting to reformulate set theory extensively. He would like to see CH tossed out and set theory recast as 'stochastic set theory', as he puts it. The notion of a *random variable* needs to be included in the fundamentals of the revised theory, not a notion defined in measure theory terms, as it currently is. Among other things, he would eliminate the power set axiom: 'What mathematics really needs, for each set X, is not the huge set 2^X but the set of sequences $X^{\mathbb{N}}$ in X' (Mumford 2000: 208).

It's easy to see why Mumford might be interested in Freiling's work – it highlights randomness. He calls the darts 'real random variables' and sharply distinguishes them from the standard mathematical notion. His interests are far from the mainstream of mathematics, including mainstream foundational work in set theory. The upshot is that most people will remain uninterested in Freiling's refutation of CH, unless they can be convinced that thought experiments involving random dart tosses can actually yield striking mathematical results that are both original and reliable.

But first, we need to backup considerably and set the stage properly.

What is the Continuum Hypothesis?

The continuum hypothesis was number one on Hilbert's famous list of problems, most of which have now been settled. There are three ways of resolving a problem such as CH: prove it true; prove it false; prove it undecidable. CH, unfortunately, is the last of these. Before dealing with undecidability, let's quickly review developments in set theory up to that point.

The natural numbers, also known as the counting numbers, are in the set $N = \{0, 1, 2, 3, \ldots\}$. The size of this set, its *cardinality*, is infinite. Symbolically, $|N| = \aleph_0$. What about other infinite sets, such as the set of even numbers, E? How big is it? Two sets have the same cardinality if and only if there is a one–one, onto function between them. Such a map exists between E and N. For instance: $1 \leftrightarrow 2, 2 \leftrightarrow 4, 3 \leftrightarrow 6, 4 \leftrightarrow 8, \ldots, n \leftrightarrow 2n, \ldots$ Thus, $|E| = |N|$. The set of rational numbers, the fractions, Q, turns out to be the same size, as well. So, we have $|E| = |N| = |Q| = \aleph_0$. We might be tempted to think that all infinite sets are the same size, but this famously turned out to be not so. The set of real numbers, R, also known as the continuum, the set of points on the line, is larger. This was proven by Cantor and is surely one of the greatest mathematical results of all time.

To prove that $|R| > \aleph_0$, we need to show two things. First, we need to show that $|R|$ is at least as big as $|N|$. This is easy, since N is a proper subset of R. So there must be at least as many members of R as of N. The second thing to show is that there is no one–one mapping between N and R. That would show that they can't be the same size. Putting these two facts together gives us Cantor's spectacular result that the real numbers are more numerous than the natural numbers.

Cantor's proof that there is no one–one correspondence is the appropriately named *diagonal argument*. We begin by assuming that there is a one–one, onto map between N and R. In fact, we can even focus on just the points in the interval $[0, 1]$. Perhaps the one–one correspondence looks something like this:

$$0 \leftrightarrow .88491625 \ldots$$
$$1 \leftrightarrow .12548179 \ldots$$
$$2 \leftrightarrow .39271254 \ldots$$
$$3 \leftrightarrow .56469848 \ldots$$
$$\vdots$$

Now let's construct a number r according to the following rule: Pick the first number in the first decimal place and change it, say by lowering the digit by 1. Thus, $r = .7 \ldots$ so far. Now take the second number in the second decimal place and change it in the same way. Then the third number in the third place, and so on. Thus, we have $r = .7015 \ldots$

Since r is a real number in the set $[0, 1]$, it should be on the list, because the list was assumed to contain *all* real numbers between zero and one. However, r cannot be on the list. It cannot be the first number on the list, since it differs in at least one decimal place, namely, the first. Similarly, it cannot be the second number on the list, since it differs in at least the second decimal place. In general, it cannot be the nth number on the list, since it differs in the nth decimal place. Thus, r is not anywhere on the list. No matter what mapping we choose between **N** and [0, 1], there will always be some number defined in the same diagonal way that r is defined, so there will always be something left off the list. There are more real numbers, or equivalently, points on a line, than there are counting numbers, which is why sets the size of **R** or greater are called uncountable. Thus, there can't be a one–one, onto function between **R** and **N**, so |**R**| must be larger than |**N**|.

The proof just presented is a special case of Cantor's theorem. It may be useful to present the full version.

Cantor's theorem: For any set, S, the cardinality of the power set, $\wp(S)$ (the set of all subsets of S), is greater than the cardinality of S. In symbols, $|S| < |\wp(S)|$.

Proof. There is a natural one–one mapping from S *into* $\wp(S)$, namely, each x maps onto its singleton, $\{x\}$. This shows that the power set is at least as big, and possibly bigger. The next step is to show that they can't be the same size, which we will do by means of a *reductio ad absurdum* argument.

Assume that there is a function f that is a one–one, onto map from S to $\wp(S)$. Let A be defined as the set of elements in S that are not members of the corresponding set in $\wp(S)$. In symbols, A = $\{x \in S: x \notin f(x)\}$. For example, if $f(a) = \{a\}$, then $a \in f(a)$, so $a \notin$ A. On the other hand, if $f(b) = \{c\}$, then $b \notin f(b)$, so $b \in$ A.

Now consider the set A itself. It is a subset of S, so A $\in \wp(S)$. Since f is one–one, onto, there must be some element, x, of S that is associated with A, that is, $f(x) =$ A. Consider the element x. Is it an element of A? If it is, then, by the definition of A, it is not. But if it is not, then by the definition of A, it is. Symbolically, $x \in$ A $\leftrightarrow x \notin$ A. This is a contradiction. So the assumption that f is a one–one onto function from S to $\wp(S)$ is false.

Power sets are bigger, but how much bigger? It's useful to have Cantor's theorem in mind, but we can revert to the special case of the real numbers when considering the question. **R** is an infinite set that is larger than **N**. But how big is it? Since each real number is an infinite decimal expansion, the set of real numbers is an infinite set of infinite numbers. This means that its cardinality is $2^{\aleph_0}$. In general, the cardinality of the power set of S is $2^{|S|}$. Cantor's theorem establishes a hierarchy of sets with infinite cardinalities: $\aleph_0 < 2^{\aleph_0} < 2^{2^{\aleph_0}} < \dots$. The interesting question he faced concerns the place of **R**, the continuum, in the hierarchy: $0 < 1 < 2 < \dots < \aleph_0 < \aleph_1 < \aleph_2 < \aleph_3 \dots$ Does $2^{\aleph_0} = \aleph_1$? Or does it equal $\aleph_2$? Or perhaps $\aleph_3$? Cantor's continuum hypothesis is the claim that |**R**| = $\aleph_1$, or equivalently, that $2^{\aleph_0} = \aleph_1$. The so-called 'generalized continuum hypothesis' is the claim that $2^{\aleph_n} = \aleph_{n+1}$. If CH is false, then |**R**| might equal $\aleph_2$, or $\aleph_{374}$, or perhaps it might be larger than $\aleph_n$, for any finite n. Then again, it could be wrong because the whole of transfinite set theory is utter rubbish. All possibilities should be kept in mind.

Though the continuum hypothesis is usually expressed in terms of transfinite cardinal numbers, these concepts are not essential to the problem. It actually arises in a very simple way in standard analysis. CH is equivalent to the claim that every set of real numbers is equivalent (i.e., there is a one–one correspondence) to a countable set of natural numbers or to the set of all real numbers.

The early twentieth century saw lots of failed attempts to prove or to refute CH. The first significant advance came in 1938 when Gödel proved that CH is consistent with the rest of set theory. He did this by providing a model based on the so-called 'constructable sets', in which all the axioms of ZFC (Zermelo–Frankel set theory with the Axiom of Choice) are true and CH is true as well. This means, of course, that CH cannot be refuted in the normal way, that is, by proving ~CH via a derivation from the axioms of ZFC.

Full independence was established in 1963 by Paul Cohen. He introduced a powerful new technique called forcing that allowed him to construct a model of set theory in which ZFC is true but CH is not. The Gödel and Cohen results together establish undecidability. CH is independent of ZFC; it cannot be proven and it cannot be refuted – at least, not in the usual way. This is how things stand.

At this point we face an interesting philosophical problem. In everyday mathematics we are happy to link *truth* with *proof*. Two philosophical camps make this explicit, formalists and constructivists, though their motivations for doing so are quite different. Since constructivists will have neither truck nor trade with Cantor's infinite sets, we will ignore them here. Formalists, on the other hand, happily embrace set theory. They typically hold the view that CH, since it has been shown to be independent, simply has no truth value — it is neither true nor false. The underlying reason for this attitude is the conviction that mathematics is a body of axioms that we accept for various reasons, but being objectively true is not one of them. To say that a mathematical proposition P is true is only to say that P can be logically derived from the accepted axioms. And to say that P is false is to say that ~P can be derived from those axioms. Neither is possible for CH, so, to the formalist-minded, it lacks any truth-value whatsoever.

Platonists, by contrast, assert that truth is distinct from proof. A proof of P does not make P true; rather it is evidence that P is true. The lack of a derivation from first principles only means that we might be forever ignorant of the truth-value of P, but P has a truth-value all the same. The instincts of any Platonist are the same as the instincts most of us have about statements in the natural sciences. We will never have evidence one way or the other that there was a T-Rex standing on the very spot I am standing exactly 75 million years ago to the second. Nevertheless, most of us believe that the claim is true or it is false. Being able to prove or refute, it has nothing to do with its truth or falsity. The Platonist attitude to CH is the same. It really is true, or it really is false, even if we cannot prove which.

Besides the general considerations stemming from Platonism, there are additional reasons for thinking CH has a definite truth-value, reasons which are motivated by considerations drawn from the details of set theory. Here (for those with some familiarity with set theory, others may ignore it) is one that is interesting, even if of only limited plausibility. Imagine working one's way through the ordinals. Each time you pass an ordinal β, pick out a real number r to associate with it, r_β. Since the set of real numbers, **R**, is just a set while the ordinals, *Ord*, is a proper class, we will certainly run out of reals before we run out of ordinals.[1] Thus, for some ordinal α, $\{r_\beta: \beta \in \alpha\}$ will exhaust **R**. It seems then that $|\mathbf{R}|$ is the least α that can be listed as $\{r_\beta: \beta \in \alpha\}$. This means that α is some cardinal number a, so we can conclude that the continuum has some cardinal number, $\aleph_a$. Of course, if $a = 1$, then CH is true; if $a \neq 1$, then CH is false. But it must be one or the other; it does have a truth value.

I include this argument for those who might find it interesting. It is not, of course, a proof, merely a suggestive consideration proposed by some set theorists. I doubt that it has the same plausibility as the vastly more plausible Platonist claim that every proposition has a truth value, whether known or not, in virtue of the independently existing realm of sets.

Though Platonists distinguish between proof and truth, they are also more inclined to entertain other types of evidence. In fact, the two go together. Since proof is not a criterion of truth, but merely a form of evidence of truth, there is a natural inclination to entertain other forms of justification. After all, the axioms themselves can't be derived, but must be thought true for some other reason. As the rest of this book attests, I take a liberal Platonist attitude here and consider mathematics to be like the natural sciences where some of the most important discoveries, such as the microscope, for instance, initiated new methods of generating evidence. Consequently, a thought experiment involving dart throwing might, at least in principle, provide evidence for the truth or the falsity of CH.

How Could We Determine the Truth of CH?

Kurt Gödel likened the epistemology of mathematics to the epistemology of the natural sciences in two important respects. First, we have intuitions or mathematical perceptions. These are the counterpart of sense perceptions of the physical world. We see that the sky is blue and that the white streak in a cloud chamber has such and such a shape. We similarly see that $2 + 3 = 5$ and that the set of even numbers is a proper subset of the natural numbers. Speaking metaphorically, we see some things with the mind's eye.

Second, we evaluate (some) mathematical axioms on the basis of their consequences, especially the consequences that we can intuit, just as we evaluate theories in physics or biology on the basis of their empirical consequences. No one can see atoms or subatomic particles, but we do see line spectra and streaks in a cloud chamber. No one can see species evolve, but we can see fossils and the geographical distribution of species with differing characteristics. What we can see is evidence for theories about things we can't see. Similarly, says Gödel, intuitions are indirect evidence for axioms in mathematics.

On Gödel's view, mathematics is fallible for a number of reasons. We can have faulty intuitions, just as we can make mistakes in our sense perceptions. Moreover, false premises can have true consequences, so the testing of axioms is not foolproof either. Many people dislike the idea of giving up certainty in mathematics; perhaps they expect axioms to be 'self-evident' truths. Others will utterly oppose the idea of intuitions, fallible or not. In what are perhaps the three most famous and most often quoted passages in all of Gödel's works, he asserts the two key ingredients of Platonism: the ontology of realism and the epistemology of intuitions. He also notes the possibility of discovering new axioms that could settle old questions, such as CH. I quoted this passage before, but it is worth repeating.

> ... despite their remoteness from sense experience, we do have something like a perception also of the objects of set theory, as is seen from the fact that the axioms force themselves upon us as being true. I don't see any reason why we should have any less confidence in this kind of perception, i.e., in mathematical intuition, than in sense perception, which induces us to build up physical theories and to expect that future sense perceptions will agree with them and, moreover, to believe that a question not decidable now has meaning and may be decided in the future. The set-theoretical paradoxes are hardly more troublesome for mathematics than deceptions of the senses are for physics ... [N]ew mathematical intuitions leading to a decision of such problems as Cantor's continuum hypothesis are perfectly possible ...
>
> (Gödel 1947/64: 484)

I take Gödel's various remarks to assert a number of important things, including: mathematical objects exist independently from us; we can perceive or intuit some of them (though not all); our perceptions or intuitions are fallible (similar to our fallible sense perception of physical objects); we conjecture mathematical theories or adopt axioms on the basis of suggestive intuitions (as physical theories are conjectured on the basis of suggestive sense perception); these theories typically go well beyond the intuitions themselves, but are tested by them (just as physical theories go beyond empirical observations but are tested by them); and in the future we might have striking new intuitions that could lead to new axioms that would settle some of today's outstanding questions. Though sketchy, these are the typical ingredients of modern mathematical Platonism. The only one I want to focus on is perception, intuition, or seeing with the mind's eye.

Gödel, as I just mentioned, took intuitions to be the counterparts of ordinary sense perception. Just as we can see some physical objects (trees, dogs, rocks, the moon), so we can intuit some mathematical entities. And just as we can see that grass is green and the moon is full, so we can intuit that some mathematical propositions are true. These perceptual facts will play a big role in deciding which propositions to accept or reject when they cannot be directly evaluated perceptually.

Kreisel's Analogy

George Kreisel (1967, 1971) shed considerable light on CH by rejecting a popular analogy that had sprung up. Following the Gödel–Cohen independence proof, it was sometimes said that CH is similar to the parallel postulate of Euclidean geometry and that there could be alternative set theories in the same way there are non-Euclidean geometries (Cohen and Hersh, 'Non-Cantorian Set Theory', 1967). Kreisel pointed out a crucial difference between the two cases. CH is only independent when we restrict ourselves to first-order set theory. It is decidable in second-order set theory. The parallel postulate, by contrast, is absolutely independent of the other postulates. This, of course, is happy news for any Platonist, for it means that CH has a definite truth value, though we still don't know what it is.

Kreisel offered a much better analogy. He took the proof of independence of CH to be like the proof that one cannot trisect an arbitrary angle with straight edge and compass. Of course, an arbitrary angle has three equal parts, but we cannot determine what they are with the impoverished method of straight edge and compass. We might, however, be able to trisect some other way. At this point, I want to underscore Kreisel's analogy with a pair of picture proofs that will make the construction of trisections perfectly evident.

In the first of these, imagine that we confine ourselves to straight edge and compass, but we allow ourselves infinitely many operations. Then we could construct a trisection. (If Figure 11.1 alone is not evident, note the infinite series that accompanies it on p. 49, where I first used this diagram and the next.)

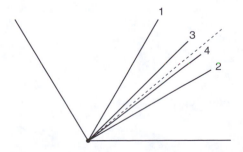

Figure 11.1

In Figure 11.2, it's even more obvious. (In case you're tempted to think that mechanical devices such as this can't be accurate and so it can't really be a tri-sector, just remember that a real straight edge is in fact a bit wobbly, a compass gives a bit under pressure, pencil lines have some finite thickness, and so on. Theorems involving these techniques are all idealizations.)

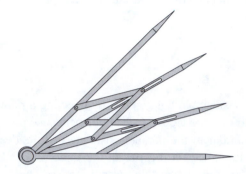

Figure 11.2

The two morals I want to draw from Kreisel's analogy will be obvious. First, CH does indeed have a truth value, and second, we might be able to determine it with a picture or thought experiment. I hope this is sufficient groundwork that we can now turn back to the refutation of CH. The assumptions of a rather general and philosophical nature we will chiefly rely upon are:

1. CH has a truth-value, even though it is independent of the rest of set theory.
2. Potential axioms and other mathematical propositions, such as CH and ~CH, could be justified in ways that are quite different from traditional proofs.
3. Thought experiments and pictorial reasoning could justify mathematical propositions by generating new intuitions.

Now to Frieling's remarkable argument.

Freiling's Refutation of CH

I'll breakdown the argument into several steps.

ZFC and well ordering

We shall take ZFC for granted as well as an important consequence of it, the so-called 'well ordering principle'. It says that any set can be well ordered, that is, can be ordered in such a way that every subset has a first element. The usual ordering, <, on the natural numbers is also a well ordering of the natural numbers. Pick any subset, say, {63972, 47, 36, 82}; it has a first element, namely 36. But unlike the natural numbers, the usual ordering on the real numbers, <, is not a well ordering. The subset $(0, 1) = \{x: 0 < x < 1\}$, for instance, does not have a first element. Nevertheless, the well ordering principle guarantees that the real numbers can be well ordered by some relation, <, even though no one has yet found such a well ordering.

Darts, points, real numbers

Freiling's thought experiment involves throwing darts at a line to select real numbers. There are important assumptions involved in this. One of these is that the line consists of pre-existing points. Aristotle, by contrast, thought that points could be constructed, say, by throwing darts, but those points do not *already* exist on the line. If Aristotle is right, then Freiling's argument would certainly not work; so the assumption of pre-existing points is crucial. And so is the assumption that points correspond exactly to real numbers. These are assumptions that almost everyone would make today, thanks, no doubt, to the success of Descartes's analytic geometry and all that rests upon it.

In addition, there are idealizations galore. Any physical line is certainly not continuous. The world is discrete in a number of respects, possibly even space itself. So it's some kind of idealized line at which we're throwing darts. Moreover, no one believes for a moment that we could identify specific real number with a dart's location. That, too, is idealized. All of which go to show that this is essentially a thought experiment.

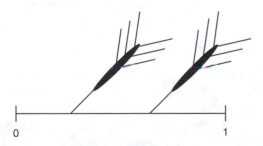

Figure 11.3

The thought experiment

Now to the visual part of the thought experiment. Imagine throwing darts at the real line, specifically at the interval [0,1]. Two darts are thrown and they are independent of one another. The purpose is to select two random numbers, *p* and q. If you fear the darts thrown one after the other might not be independent, then imagine two dart throwers who are screened off from one another and who both throw at the target line on the count of 'one, two, three, go', or separate them and have them toss at different copies of [0,1].

There are three important things to notice in the thought experiment: a pair of real numbers are selected (1) randomly, (2) independently, and (3) symmetrically. Let's flesh these out a bit.

Random variables, independence, and symmetry

The concept of random variable at work here is not the mathematical concept found in measure theory. Standard mathematical definitions of random variables go something like this: A *random variable* is a measurable function from a probability space into a measurable space. It's easy to understand by means of an example. Imagine rolling two dice. We can number each roll: roll 1, roll 2, roll 3, and so on. The outcome of a roll will be some number between 2 and 12. You can think of a roll as an experiment and the number that turns up on the dice as the result. The random variable is the function X with domain {roll 1, roll 2, . . . } and range {2, 3, . . . , 12}. Tossing a dart is thus a random variable with outcome in [0,1]. These are 'real random variables', says Mumford (2000).

The two real numbers are picked independently. This is obvious, since the two dart throws have no influence on one another. This means that a prediction based on either throw cannot be dismissed in the way we might dismiss someone who said of a licence number on a passing car: 'Wow, there was only a one in a million chance of that happening.' We're rightly impressed only if the number is fixed independently of the outcome (i.e. predicted before the result is known). The independence and randomness of the darts guarantees the symmetry of the throws. Consequently, either dart could be considered the first throw that fixes the set of real numbers that are earlier in the well ordering.

Cardinals, ordinals, and initial segments

There's an important fact from set theory that we shall use in the following argument. I'll explain it now. The ordinal numbers are ordered (indeed, well ordered): $0 < 1 < 2 < \ldots < \omega < \omega + 1 < \omega + 2 < \ldots < \omega2 < \omega2 + 1 < \ldots < \omega3 < \ldots < \omega^2 < \ldots < \omega^\omega < \ldots < \epsilon_0 \ldots$ Ordinal numbers have the very important property that any two are isomorphic or that one is isomorphic to an initial segment of the other. An *initial segment* of an ordered set is simply a proper subset. If S is an initial segment of a well-ordered set W, then there is an $a \in W$ such that $S = \{x: x < a\}$. Thus, the number $3 = \{0, 1, 2\}$ is an initial segment of the ordinal number 4 (and also of 5, and 6, and 7, . . .). For any number n, the set $\{0, 1, 2, \ldots, n\}$ is an initial segment of ω, the ordinal number that is the set of all the natural numbers. An important theorem states: No well-ordered set is isomorphic to one of its initial segments.

Cardinal numbers are defined in terms of ordinals. In the finite case, they are simply identified. Thus the ordinal 27 is the same as the cardinal 27.[2] In the infinite case, things are a bit more complicated. There are many ordinals that are the same size in the sense of there being a one–one mapping between them. For instance, $\omega, \omega + 1, \omega + 2, \ldots, \omega2, \omega2 + 1, \omega2 + 2, \ldots, \omega3, \omega3 + 1, \ldots, \omega^2, \omega^2 + 1, \ldots$ can all be put in one–one correspondence with each other. However, there is a rather natural definition: a *cardinal number* is the least of a set of equivalent ordinals. Thus, $\aleph_0$ is identified with ω.

There is a consequence of this that you may have noticed. Since any initial segment of an ordinal will be smaller than the ordinal itself and since a cardinal number is identified with the least equivalent ordinal, it follows that the cardinality of an initial segment must be smaller than the cardinality of the set we start with. In some cases this is quite intuitive. If we start with the set ω and pick any initial segment, then we will have picked a set with only finitely many members. It's obvious in this example, but it holds in general. Consequently, if we started with a set of cardinality $\aleph_1$ and picked an initial segment, the cardinality of the initial segment would be countable, that is, it would be $\aleph_0$ or finite. We will use this fact in the argument below.

Measure and probability

In any finite case the concept of probability is readily understood. The infinite case is tricky. If we throw a pair of dice, there are 36 possible results. Representing these as a pair (first die, second die), we have a so-called 'probability space', $\{(1,1), (1,2), (2,1), \ldots, (6,6)\}$, with 36 distinct outcomes. Assuming these are fair dice, then each outcome is as likely as any other. The probability of getting the result 2 is 1/36, since there is only one way this could happen, namely, by rolling (1,1). There are three ways of getting the outcome 4, namely, (1,3), (3,1), and (2,2). Thus, the probability of getting the result 4 is 3/36 = 1/12. The probability of getting a result that is an even number is one half, of getting a result between 2 and 12 is one, and of getting the result 13 is zero. This is all perfectly straight forward, but things are not so easy in the infinite case.

Toss a dart at the line segment [0,1]. You might hit 1/5, $2/\pi$, or $e/3$, and so on. But what is the chance of hitting any one of these? One chance in infinitely many, which means the probability is zero. Surprisingly, events with a zero chance of happing can nevertheless actually happen. This is weird, but not logically absurd. One might think making sense of the infinite case is hopeless to manage, but not so. A branch of analysis known as measure theory has come to the rescue.

Measure theory, or more specifically, the theory of Lebesgue measures, gives us a way of assigning a measure to a huge number of different sets. The measure of a set of points that is an interval is simply its length. Thus, the length of the line between segment between 7 and 13 is equal to $13 - 7 = 6$. Symbolically, $\mu([7,13]) = 6$ (where μ is the Greek letter mu, commonly used for this purpose).

Obviously, $\mu([0,1]) = 1$. The second principle of measure theory says that if a set S equals the union of a countable number of disjoint subsets, that is, $S = s_1 \cup s_2 \cup s_3 \cup \ldots$, each of which is measurable, then $\mu(S) = \mu(s_1) + \mu(s_2) + \mu(s_3) + \ldots$. To take a simple case, if S is the set consisting in the segments [0, 1/2], [3/4, 7/8], [9/10, 1], then $\mu(S) = 1/2 + 1/4 + 1/10 = 17/20$.

Probability is easily understood in these terms. In throwing a dart at the line [0,1], the chance of hitting the segment [0,1/2] is clearly 1/2, and so on. Thus, the probability of landing in S is equal to the measure of S. So far, so good. But what about the probability of hitting a rational number? What is the measure of the set of fractions in [0,1]? This is a set of points that is distributed throughout [0,1], but is certainly not an interval. I'll use $\mathbf{Q}_1$, $\mathbf{R}_1$, and $\mathbf{I}_1$ to be the set of rational, real, and irrational points in [0,1].

The measure of any singleton set is zero; that is, a single point, a, has no length, so $\mu(\{a\}) = 0$. As we already know, $\mathbf{Q}_1$ is countable. A theorem of measure theory says the following: If S is the countable union of sets of measure zero, then S is also of measure zero. Since $\mathbf{Q}_1$ is the union of countably many sets of single points, each having measure zero, it follows by the theorem that $\mathbf{Q}_1$ also has measure zero. What about the irrational points? We already know that $\mathbf{I}_1$ is uncountable, so the theorem does not apply to it. We can, however, easily determine its measure. Since $\mathbf{R}_1 = \mathbf{Q}_1 \cup \mathbf{I}_1$, it follows that $\mu(\mathbf{R}_1) = \mu(\mathbf{I}_1) + \mu(\mathbf{Q}_1)$. And since $\mu(\mathbf{R}_1) = 1$ and $\mu(\mathbf{Q}_1) = 0$, it follows that $\mu(\mathbf{I}_1) = 1$. The real line is overwhelmingly dominated by the irrational numbers. In terms of probability, the chance of hitting an irrational number with a dart is one.

Measure theory allows us to talk about the measure of some pretty strange sets, not just rational and irrational ones. We'll encounter one of these strange sets momentarily. The crucial thing to remember is that the measure of any countable subset of [0,1] is zero and hence, the probability of hitting any member of that set is also zero.

The argument

1. We assume ZFC and we further assume (with the aim of generating an absurdity) that CH is true.
2. We toss two darts at the real interval [0,1] in order to pick out two real numbers.
3. The points on the line can be well ordered so that for each $q \in [0, 1]$, the set $\{p \in [0, 1]: p < q\}$ is countable. (Note that $<$ is the well ordering relation, not the usual less than, $<$,.) The well ordering is guaranteed by ZFC. The fact that the set is countable stems from the nature of a well ordering of any set that has cardinality $\aleph_1$, as was explained above. To repeat, a cardinal number is defined as the least of all the equivalent ordinals, so the initial segment defined by q must be a smaller cardinal than the cardinal number of [0,1],

which, given our assumption CH, is $\aleph_1$, the first uncountable number. Thus, $\{p \in [0, 1]: p < q\}$ must be countable.

4.   We shall call the set of elements that are earlier than the point p in the well ordering S_p. Suppose the first throw hits point p and the second hits q. Either $p < q$, or vice versa; we'll assume the first. Thus, $p \in S_q$. Note that, for the reason stated immediately above, S_q is a countable.

5.   Since the two throws are independent of one another, we can say the throw landing on q defines or fixes the set S_q in a way that is independent of the throw that picks out p.

6.   The measure of any countable set is zero, thus, $\mu(S_q) = 0$. So the probability of landing on a point in S_q is also zero.

7.   By the same line of reasoning, we can define a set S_p of points that precede p in the well ordering and that also has measure $\mu(S_p) = 0$.

8.   One of the two darts must land in a set defined by the other dart, even though the probability of doing so is zero. While logically possible, this sort of thing is almost never the case. Yet it will happen every time there is a pair of darts thrown at the real line. This is absurd.

Conclusion: We should therefore abandon the initial assumption, CH, since it leads to this absurdity. Thus, CH is refuted and so the number of points on the line is greater than $\aleph_1$.

What Might the Continuum Be?

If the cardinality of the continuum is $\aleph_2$ or greater, the argument (at least as set out here so far) would not work, since the set of points S_q earlier in the well ordering need not be countable, and so would not automatically lead to a zero probability of hitting a point in it.

A large number of set theorists, including Gödel, Cohen, Woodin, and others already believe that CH is false. Cohen is particularly adamant.

A point of view which the author feels may eventually come to be accepted is that CH is *obviously* false. The main reason one accepts the Axiom of Infinity is probably that we feel absurd to think that the process of adding only one set at the time can exhaust the entire universe. Similarly with the higher axioms of infinity. Now $\aleph_1$ is the set of countable ordinals and this is merely a special and the simplest way of generating a higher cardinal. The set $\wp(\omega)$ is, in contrast, generated by a totally new and more powerful principle, namely the Power Set Axiom. It is unreasonable to expect that any description of a larger cardinal which attempts to build up that cardinal from ideas deriving from the Replacement Axiom could ever reach $\wp(\omega)$. Thus $\wp(\omega)$ is greater than $\aleph_n$, $\aleph_\omega$, $\aleph_{\omega\omega}$, etc. This point of view regards $\wp(\omega)$ as an incredibly rich set given to us by one bold new axiom, which can never be approached by any piecemeal process of construction.

(Cohen 1966: 151)

Many today hold the view that $|\mathbf{R}| = \aleph_2$, so the argument here harmonizes with this.[3] However, Freiling extends the initial thought experiment, plausibly arguing that the continuum is not $\aleph_2$, nor $\aleph_3$, nor $\aleph_4$, and so on. If we throw a third dart, we are unlikely to land in either set defined by the first two darts. This yields another axiom similar to the symmetry axiom described in the appendix, which in turn leads to the theorem that the continuum must be greater than $\aleph_2$. A fourth dart justifies another symmetry axiom and the consequent theorem that the continuum must be greater than $\aleph_3$. Continuing in this manner, we can show that the continuum is greater than any finite aleph, that is, $2^{\aleph_0} > \aleph_n$, for any finite n.

Freiling uses the dart method to argue for a number of other results that I won't describe here. For instance, he casts doubt on the Axiom of Choice, the Well-ordering theorem, Martin's Axiom, and many others. I won't try to evaluate these additional arguments, but instead direct readers to his paper (Freiling 1986).

Two Objections

The refutation of CH made use of a principle to the effect that when picking out an initial segment, we end up with a set of lower cardinality. We can use this fact to get apparently paradoxical results from smaller well-ordered sets. For instance, pick a pair of natural numbers at random. Let them be m and n. Suppose m is chosen at random. What is the probability that n is less than m? It's zero. Similarly, the probability that n is less than m is also zero. Does this refute the view that the cardinality of the natural numbers is $\aleph_0$? The answer is No, but we should reject the claim that this argument is parallel to the Freiling's.

The conclusion this argument actually justifies is that we cannot talk about the probability of picking natural numbers at random. We can pick them at random out of the bounded set $\{0, 1, 2, \ldots, n\}$, but not out of the set $\mathbf{N}$. This is where the dart throwing thought experiment plays a crucial role. We cannot throw darts at the natural numbers in the same way we can throw them at the reals between $[0,1]$. We could crowd the natural numbers onto the interval $[0, 1]$, but the chance of hitting any one of them would be zero. Or we might try the suggestion that the dart picks out the nearest natural number. This would be a kind of randomness, but another problem arises. It turns out that how we locate the natural numbers on the real line matters. We might put the number 1 at the point 1/2, the number 2 at the point 3/4, 3 at the point 7/8, and so on. Now the probability of hitting a real number close to the associated natural number 1 will be much greater than hitting one close to 10.

In short, while the thought experiment allows us to pick random real numbers, there is, I strongly suspect, no such process for picking natural numbers at random out of **N**. If we stay completely within the mathematical realm, the idea of picking numbers at random can lead to paradox. But when we move to the dart-throwing thought experiment we introduce something genuinely new. This shows, I think, that the thought experiment is essential to the argument. It is not a mere heuristic device that could, in principle, be eliminated.

A second objection concerns the reliability of the intuitions in the thought experiment. I said at the outset that Freiling's work was ignored. This isn't completely true. Every now and then there is a flurry of activity on the Internet. A discussion on FOM (a Foundations of Mathematics discussion list) was largely critical of Freiling for the simple reason that intuitions are untrustworthy. The sentiments of two discussants are typical.

> Freiling's argument depends on assuming that the concept of randomness/probability/measure applies to certain 'weird' sets associated with a well-ordering of the reals. We've all been indoctrinated in school about how the axiom of choice lets us construct non-measurable sets, so I don't see why we should believe that the particular weird sets in Freiling's argument should be measurable.
>
> (Timothy Chow)

> We are importing our intuitions about ordinary physical objects into a context where they make no sense. Partitioning a ball (as in the Banach-Tarski 'paradox') has nothing to do with 'cutting a ball into pieces' in the ordinary physical sense. In the case of Freiling's argument, what sense does it make to say 'I threw a dart at the wall, and hit a point with rational coordinates.' None whatsoever!
>
> (Alasdair Urquhart)[4]

Intuitions can indeed lead us astray. Thought experiments are fallible. However, ordinary experience is also mistaken from time to time, but we would be fools to toss out observation as a source of knowledge simply because we sometimes suffer illusions. The advance of science includes a better understanding of the process of observation itself. We have learned some of the optical properties of the atmosphere leading us to distinguish the true position of Mars from its apparent position. We have learned and we continue to learn about how observation in a microscope works, allowing us to distinguish what is 'really' there from what is an artifact of the observing process, such as damage done by staining. We should take the same attitude towards mathematical intuitions. Accept them in principle, but proceed cautiously and critically.

The Banach–Tarski paradox, for instance, is very counter-intuitive. It tells us that we can decompose a basketball, for example, into a finite number of parts and re-compose it to make another ball the size of the earth. Needless to say, 'decompose' and 're-compose' are not physical operations in any sense. Even so, it is quite bizarre. If we knew nothing about the Axiom of Choice except that it had this consequence, we would almost certainly toss it out. But the axiom is known to solve many outstanding problems that couldn't be solved otherwise; it is extremely fruitful, in Gödel's important sense of unification and problem solving. The axiom, in the infinite case, is supported by analogy with the obviously true finite case. Moreover, if one is already a Platonist, then the idea of arbitrary functions and arbitrary sets (which is essential to the axiom) is completely natural. These considerations have lead us to accept the axiom as a part of standard set theory. Today, there is little or no concern with the Axiom of Choice and almost everyone cheerfully uses it. The moral drawn is that we should accept the Banach–Tarski result and overrule our intuitions in that specific case, just as we have accepted Copernicanism and come to believe that the earth moves, not the sun, in spite of appearances to the contrary. Our working rule should be to accept an intuition, unless it's shown to be faulty. The degree of acceptance should be linked to the strength of the intuition and to its integration into our whole system of beliefs. So, a powerful intuition that refuted CH should be accepted, unless there is some reason to think the intuition is tainted or unreliable. Is there such a reason?

Timothy Chow, in the passage quoted above, says Freiling's example might involve non-measurable sets, and we know from Banach–Tarski type examples that our intuitions of non-measurable sets are not to be trusted. But, as a matter of fact, we do not know that non-measurable sets are involved in this case. On the contrary, the opposite view seems justified. The set S_q is countable, and any such set has measure zero. So, the set we care about is indeed measurable. Until demonstrated otherwise, I see no reason for Chow's scepticism.

Alasdair Urquhart says throwing darts has nothing to do with picking out numbers and that we are fooled by the physical analogy. In further private discussion he remarks: "My point is that by dressing up mathematical propositions [e.g., CH] in 'physical' language, you can make them sound completely implausible. For example, it wouldn't be hard to take a perfectly ordinary theorem of ZF and make it sound completely implausible by making up some 'physical' story."

No doubt, he's right. Imagine a simple example: the infinite sequence of alternating positive and negative ones: $1, -1, 1, -1, 1, -1, \ldots$ This sequence does not converge. But we can think of it as a physical sequence of turning on $(+1)$ and off (-1) a light switch. Imagine doing it at the rate of one second for the fist switch from on to off, then half a second for the next switch, and so on. In two seconds we would have switched it infinitely many times. That in itself is very unphysical, but there's worse to come. After two seconds we have finished running through the series. The light should be on or off, which means the series does converge after all. The imagined physical process seems to (mistakenly) overthrow the mathematical theorem.

The moral I would like to draw from this, however, is not the same as Urquhart draws. He implicitly suggests that thought experiments of any sort, whether they involve darts or light switches or whatever, should be barred. He's right to be concerned, but goes too far and eliminates too much. If he were right, the same evaluation would have to be made of a great many thought experiments in physics. Think how unphysical they often are: Einstein chasing a light beam, Maxwell's demon, Newton's bucket in an otherwise empty universe, and so on. Calling them unphysical is probably understating things; they are anti-physical in that they actually contradict known physical principles. These thought experiments are far removed from the actual physical world, just as picking out real numbers with a dart is far removed from the physical world. However, the crucial thing about thought experiments — in physics or in mathematics — is that they clarify and illuminate conceptual matters. This is what Einstein, Maxwell, and Newton did, in spite of being highly unphysical. And this, I think, is what Freiling has done.[5]

Of course, thought experiments are fallible and further analysis can show any one of them to be faulty. This has been the fate of many of the great ones in the past and it may be the fate of Freiling's. But for now it seems a fair bet and there is no good reason to resist it, except for the normal prudence with which we sensibly approach something far from the ordinary.

What Did the Thought Experiment Contribute?

Freiling's dart thought experiment (TE) is essential to the argument, but it is far from clear just what TE actually contributed. Nevertheless, it certainly contributed something. The argument for this is simple and straightforward. ZFC does not imply ~CH; however, ZFC + TE does imply ~CH. Thus, TE is essential to the argument, not a mere heuristic aide.

The standard mathematical concepts of a random variable, independence, and so on are concepts that are defined (or definable) inside set theory. Since CH is independent of set theory, at least some of the concepts and principles involving them that are being used in Freiling's argument must be different from anything in standard mathematics. This is a very important matter. The thought experiment is not a mere heuristic device that helps the imagination but is eliminable in principle. In fact, it is essential to the result.[6]

The thought experiment is actually providing something new. But what is it? The obvious answer as far as mathematics is concerned is *Freiling's Symmetry Axiom*: $(\forall f: \mathbf{R} \to \mathbf{R}_{\aleph_0})(\exists x)(\exists y) \, y \notin f(x) \, \& \, x \notin f(y)$ (see appendix for a discussion). But now the question becomes: How does it justify that? My answer is Platonistic. The thought experiment provides an access to the abstract realm where we can 'see' with the mind's eye. It does this by generating concepts of randomness, symmetry, and independence that are different from existing versions of those concepts. And these new concepts, together with existing principles, allow us to derive the new result.

Two Morals

The first moral, of course, is that CH is false. Naturally, we shouldn't be as confident of this as we would be of any simple theorem proved in the standard way. Nevertheless, the result seems as solid as many of the things we believe about the physical world.

The second moral may, in the long run, prove the more important. Picture proofs and thought experiments are a potential source of mathematical knowledge that is largely untapped. They ought to be explored and exploited. This is a resource that has hitherto been confined to the role of heuristic device and a psychological aide, but nothing more. On the contrary, as I have been arguing throughout this book, a great deal more is possible. Only the hopelessly unimaginative will accept the view that some mathematical problems are truly unsolvable. They may be unsolvable by existing methods, but there is no reason to tie out hands with such impoverished tools. I doubt that thought experiments will solve all problems, but they might solve a few that cannot be solved otherwise. Neither God nor Gauss has forbidden their use. And even if they had, we should thumb our noses and sail on.

Appendix: Freiling's Version

Freiling (1986) is slightly different than the version I have given above. I'll reproduce his actual argument here. He assumes the following four 'self-evident philosophical principles':

1. Choosing reals at random is a physical reality, or at least an intuition mathematics should embrace to the extent possible.
2. A fixed Lebesgue measure zero set predictably will not be hit by a random dart.
3. If an accurate Yes–No prediction can *always* be made after a preliminary event takes place (e.g., the first dart is thrown) and, no matter what the outcome of that event, the prediction is *always* the same, then the prediction is also in some sense accurate before the preliminary event.
4. The real number line cannot tell the order of the darts.

(Freiling 1986, 199)

Freiling's argument runs as follows: We throw two darts, one after the other, at the real line [0,1]. Let $f\colon \mathbf{R} \to \mathbf{R}_{\aleph_0}$ be a function that assigns a countable set of real numbers to each real. The number hit by the second dart will (with probability one) *not* be in the countable set assigned to the number hit by the first dart. The situation is symmetrical; the order of throwing is irrelevant. Thus, we can say that the number hit by the first dart will not be in the set assigned to the second. This leads to the following intuitive principle that I'll call *Freiling's Symmetry Axiom* (FSA): $(\forall f\colon \mathbf{R} \to \mathbf{R}_{\aleph_0})(\exists x)(\exists y)\; y \notin f(x) \;\&\; x \notin f(y)$

Theorem (of ZFC): FSA $\Leftrightarrow$ ~CH

Proof: ($\Rightarrow$): Assume FSA and let $<$ be a well ordering of R. The existence of a well ordering follows from the axiom of choice which we have assumed as part of ZFC. We will further assume CH which implies that the length of the well ordering is $\aleph_1$. Now let $f(x) = \{y\colon y \leqslant x\}$. Thus, $f\colon \mathbf{R} \to \mathbf{R}_{\aleph_0}$. The way cardinal numbers are defined implies that we are always bumped down a cardinality when picking a set of earlier points in a well ordering. Moreover, a well ordering is total, so if some particular $y \notin \{y\colon y \leqslant x\}$, then $x > y$. Consequently, by FSA, $(\exists x)(\exists y)\; x < y \;\&\; y > x$, which is a contradiction. Therefore, ~CH.

($\Leftarrow$): Assume that CH is false, i.e., $2^{\aleph_0} > \aleph_1$. Let $x_1, x_2, x_3, \ldots$ be an $\aleph_1$-sequence of distinct real numbers and let $f\colon \mathbf{R} \to \mathbf{R}_{\aleph_0}$. Now consider the set $A = \{x\colon (\exists \alpha < \aleph_1)\; x \in f(x_\alpha)\}$, which is the $\aleph_1$-union of countable sets. Thus, the cardinality of A is $\aleph_1$. Since $2^{\aleph_0} > \aleph_1$, $\exists y \notin A$. Thus, $(\forall \alpha < \aleph_1)\; y \notin f(x_\alpha)$. Since $f(y)$ is countable, we have $(\exists \alpha \in \aleph_1)$ $x_\alpha \notin f(y)$. Therefore, $y \notin f(x_\alpha) \;\&\; x_\alpha \notin f(y)$.

Further Reading

Set Theory, including the standard results on infinite sets, is found in numerous excellent texts. Enderton, *Elements of Set Theory*, is one among many. The independence of CH is shown in Bell, *Boolean-Valued Models and Independence Proofs in Set Theory*, and in Kunen, *Set Theory*. Of course, the articles by Freiling and Mumford are essential.

Maddy, *Naturalism in Mathematics*, especially Part I, contains a wealth of material on justifying new axioms in set theory. Tiles, *The Philosophy of Set Theory*, presents an extensive treatment that is quite accessible. For advanced readers, Woodin, 'The Continuum Hypothesis, Parts I and II', presents current work, including his own, on CH.

CHAPTER 12
Calling the Bluff

Jean Dieudonné is one of the more prominent members of the Bourbaki group[1], a highly influential collection of French mathematicians who stress axiomatics and rigour in the development and presentation of mathematics. In the preface to his *Foundations of Modern Analysis* (1969), Dieudonné urges a 'strict adherence to axiomatic methods, with no appeal whatsoever to "geometric intuition", at least in the formal proofs: a necessity which we have emphasized by *deliberately abstaining from introducing any diagram in the book*' (1969: ix *Preface*, my italics).

This attitude towards pictures, illustrations, and diagrams in mathematics is widespread and long-standing. Two centuries earlier, Dieudonné's countryman, Joseph-Louis Lagrange, one of the greatest mathematicians of all time, remarked in his *Mécanique Analytic* (1788), 'No figures will be found in this work. The methods that I set forth require neither constructions nor geometrical or mechanical arguments, but only algebraic operations, subject to a regular and uniform procedure' (1788 *Pref.*).

And sure enough, there isn't a picture to be found in either book.

The fact that there aren't any diagrams makes it look as if their point is clearly made: rigorous mathematics can and should be done *without* pictures. But as Dieudonné himself puts it in describing the background required of his readers: 'students must have a good working knowledge of classical analysis before approaching this course' (1969: x *Pref.*). And where do they get that? No doubt they typically get it from lectures with lots of diagrams and a more elementary text with lots of pictures. What this means is that Lagrange and Dieudonné and so many like-minded others – though they may not realize it – can count on their readers being already familiar with the typical diagrams that go along with the various theorems. No one is learning the intermediate value theorem, for example, completely from scratch without a visual illustration.

One of the biggest claims I've made in this book (back in Chapter 3) is that pictures *in some cases* can serve as perfectly rigorous evidence. I'm *not* claiming that this is always so, or that pictures are necessary in any particular case. But we should keep in mind that high-powered, sophisticated mathematics books which have no diagrams in them are read by people who earlier read many books which did contain pictures – and those readers haven't suddenly forgotten what they learned earlier.

Pierre Cartier, a member of the Bourbaki group, was asked why the lack of visual representation in the Bourbaki books. 'The Bourbaki were Puritans', he answered, 'and Puritans are strongly opposed to pictorial representations of truths of their faith' (Senechal 1998: 27). Cartier's amusing speculation, presumably, is tongue-in-cheek. Let's look briefly at some of the more earnest reasons that have been offered for taking a dim view of pictures. One of these regards rigour. Every now and then mathematics becomes especially concerned with the security of its foundations. In the nineteenth century rigour and the arithmetization of analysis went hand in hand. Set theory was developed in connection with this, but led to many paradoxes. The most famous of these was Russell's concerning the set of all sets which are not members of themselves. So, in the early part of this century rigour was again of great concern; consequently, visual reasoning once again took a tumble. But when we consider one of the reasons for the obsession with rigour, namely the paradoxes,[2] we notice that *not a single problem arose from visual reasoning*. The paradoxes (Russell's Paradox, the Burali–Forti Paradox concerning the set of all ordinals, Cantor's Paradox concerning the set of all cardinals, and other related paradoxes) all stem from verbal/symbolic reasoning. That is what led us astray – not pictures. The same can be said of other great crises in the history of mathematics concerning infinitesimals in the seventeenth century or incommensurable measures in Greek times. Pictures were not to blame.

Formalists and linguistic conventionalists see mathematical truth (such as it is) residing in the very language or notation of mathematics. A.J. Ayer (1936/1971), in particular, held a version of this in which all mathematical truths stem from the way we use language. '5 + 7 = 12' is true for the same reasons that 'All bachelors are unmarried males' is true – that's the way we use those terms. 'We see, then', says Ayer,

> that there is nothing mysterious about the apodictic certainty of logic and mathematics. Our knowledge that no observation can ever confute the proposition '7 + 5 = 12' depends simply on the fact that the symbolic expression '7 + 5' is synonymous with '12', just as our knowledge that every oculist is an eye-doctor depends on the fact that the symbol 'eye-doctor' is synonymous with 'oculist'. And the same explanation holds good for every other sort of a priori truth.'
>
> (Ayer 1936/1971, 113)

Ayer also had it in for pictures. He expressed his view in connection with geometry which he took to be, like arithmetic, a body of analytic truths.

> It might be objected that the use of diagrams in geometrical treatises shows that geometrical reasoning is not purely abstract and logical, but depends on our intuition [i.e. visualization] of the properties of figures. In fact, however, the use of diagrams is not essential to completely rigorous geometry. The diagrams are introduced as an aid to our reasoning. . . . It shows merely that our intellects are unequal to the task of carrying out very abstract processes of reasoning without the assistance of intuition. . . . Moreover, the appeal to intuition, though generally of psychological value, is also a source of danger to the geometer.
>
> (ibid.: 111)

Aside from being a source of error, Ayer does not give us an argument for rejecting diagrams as anything more that psychological aids. However, there may be a reason implicit in his remarks that we can ferret out. If mathematical truths rest on linguistic facts, then mathematical evidence should be related to the source of its truth, namely, to language itself, to facts about verbal/symbolic usage. Thus, considerations of language are the only source of rigorous evidence (i.e. only verbal/symbolic proofs are legitimate) and pictures are properly ruled out.

The crucial thing wrong with this argument is the premiss that mathematical truth rests uniquely on verbal/symbolic facts. If, as Platonism maintains, there is more to mathematical reality than mathematical language (which is merely an instrument to represent non-linguistic mathematical reality), then pictures might be another way to represent that reality. This doesn't establish the legitimacy of pictures, but it certainly undermines a potentially effective argument against them.

One of the most powerful objections to taking pictures seriously is that the intuition which goes along with visual reasoning gets in the way of great breakthroughs in mathematics. Hans Hahn, in an influential article called 'The Crisis in Intuition' (1956) made this case long ago using some very nice examples. Hahn's official target in this paper is Kant's doctrine of intuition but, in fact, he is also aiming at both *visualization* and *intuitive* in the more common-sense meaning of 'immediately obvious'. He contrasts intuition in all of these senses with 'analytical' and 'logical' reasoning, which he takes to be the one and only way to proceed correctly.

Hahn attempts to make his case with a number of striking examples. One of these is a celebrated result of Brouwer.[3] Imagine a map with three countries. At most of the boundary points two countries touch, but at some points all three touch (Figure 12.1).

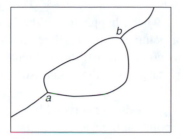

Figure 12.1 Most boundary points touch only two countries, but *a* and *b* touch all three

Pictures and common-sense intuition tell us that the special points where all three countries touch are special indeed, and that they must all be *isolated* points. Brouwer, however, showed how to construct a map in which *every boundary point touches all three countries*. Figure 12.2(a–d) gives the idea of how the construction works.

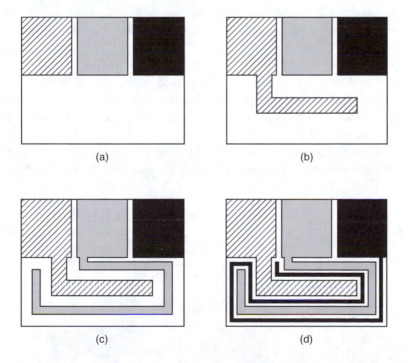

Figure 12.2 First stages in a construction in which every boundary point is common to each country

Imagine country A pushing out its boundary into the (white) unclaimed land. It takes a year to do this. Then B, not wanting to be outflanked, pushes out, too. This takes half a year. Then C pushes out, taking one quarter year. Next, it's A's turn again, and so on. After two years (a finite time) we have a situation in which *every boundary point is a point at which A, B and C meet.* The pictures help the imagination in getting a grip on what's happening in the proof but, as Hahn rightly notes, there is no hope of being able to visualize the final construction in which every boundary point is common to all three countries. It is the analytic proof (which I won't duplicate here) that establishes the result.

Another of Hahn's examples is the class of 'space-filling curves'. Imagine a square. You may start with your pencil at any point on the square and move as fast as you like (though still with finite speed) for a finite time. Can you cover every point on the square? In other words, could any curve in the square *fill* the whole square?

Hilbert's space-filling curve is generated recursively. The sequence of pictures in Figure 12.3 show the first six iterations of the construction. Imagine drawing these, each picture in half the time it took to do the previous one. If the first takes a minute, we can be done drawing infinitely many pictures in two minutes. Thus, the final curve is drawn in finite time; it is continuous, and it fills the whole square. Of course, no actual picture captures this; the sequence shown here is at best suggestive. Only the analytic proof truly establishes the result.[4]

Pictures played an important role in both Brouwer's topological result and Hilbert's space-filling curve. But the role played is the role that traditional attitudes have long assigned to pictures: psychological and heuristic – *not* evidential. In neither case do pictures provide a proof of the result; only the verbal/symbolic proof does that.

The point that Hahn makes so well with his well-chosen examples is similar to one that Descartes made long ago. If we try to imagine a 1000-sided figure

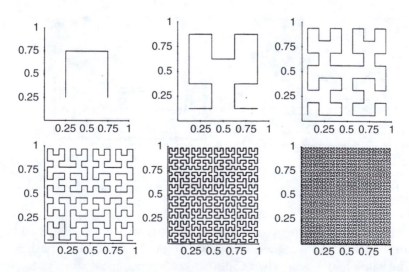

Figure 12.3 The first six iterations of Hilbert's space-filling curve

and a 1001-sided figure, we can't see the difference, at least not in the imagination. But the intellect has no trouble at all distinguishing. A picture may be worth a thousand words, but here a (number) words can distinguish among thousands of pictures.

Any reply to Hahn's argument must concede a great deal – pictures can indeed lead us astray. But his argument is not general; only some intuitions are misleading – not all. The case that Hahn raises against pictures is perhaps the most powerful that could be made. His examples seem to show an in-principle limitation to visual reasoning. In some cases, this is so. Nevertheless, this point should not be generalized, since lots of other pictures work very well. But even though these are very powerful examples, they have not been the most effective elements in the historic campaign against visual thinking. Hahn's examples mislead in not giving us the right result – as opposed to convincing us of an outright falsehood. I suspect that pictures have suffered more from the fact that they have led people into explicitly false beliefs than from any other reason.

The kind of thing I mean is well illustrated by a popular puzzle. Consider a square with sides of length 13 (Figure 12.4(a)). Divide it into four sections as indicated, then rearrange as in Figure 12.4(b). The area of the first square is $13 \times 13 = 169$, but the area of the second is $8 \times 21 = 168$. What went wrong? Where is the missing area? (As a final exercise, you might try to figure it out. Answer on p. 000)

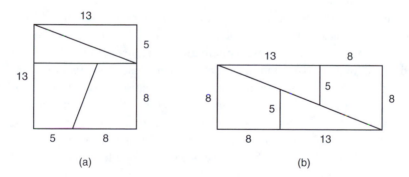

(a)   (b)

Figure 12.4 Do they have the same area?

Here's another example, less well known, but quite striking.

Draw four circles in the plane, centred at $(\pm 1, \pm 1)$, each with radius 1 (Figure 12.5(a)). Draw a fifth circle, this time at the origin, so that it touches each of the four circles. Draw a box around the four circles. It will have sides stretching from -2 to $+2$. Obviously, the inner circle is completely contained within the box. Do the same in three-dimensional space, this time drawing eight spheres centred at $(\pm 1, \pm 1, \pm 1)$ and a ninth sphere at the origin touching the other eight. Now draw a box around the eight spheres. Once again, the central sphere is completely contained within the box (Figure 12.5(b)).

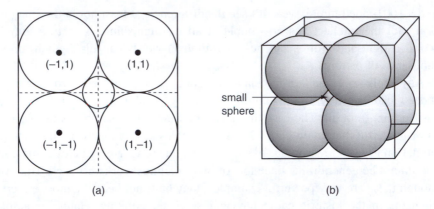

Figure 12.5 Will the small central sphere stay contained in the box?

Reflecting on these pictures, it would be perfectly reasonable to jump to the 'obvious' conclusion that the result holds in higher dimensions. Amazingly, this is not so. At ten dimensions or higher, the central sphere breaks through the *n*-dimensional box. Here's why: The distance from the origin to the centre of any sphere is $\sqrt{((\pm 1)^2 + \ldots + (\pm 1)^2)} = \sqrt{n}$. But each sphere has radius 1; thus, the radius of the central sphere is $\sqrt{n} - 1$. For $n \geq 10$, we have $\sqrt{n} - 1 > 2$. Thus, the central sphere will break through the sides of the *n*-dimensional box – a profound shock to intuition.

What is the moral to be drawn from examples such as this? The all-too-common response is to relegate pictures to *heuristic only* status and to say that they are not to be trusted as mathematical evidence. But as I repeatedly pointed out above, verbal/symbolic proofs can mislead, too. Pictures are no worse, and can even correct faulty symbolic derivations. It would be much better to consider the evidence acquired from pictures to be like the empirical evidence acquired from microscopes, bubble chambers, and other instruments of observation. These instruments can be highly misleading, too. Optical properties and staining techniques which were not understood have led microscope users to 'observe' things that are not real, but were mere artifacts of the observation process. Students of biology and astronomy spend considerable time learning about the potential pitfalls associated with optical instruments. Mathematics students could be given similar training in connection with diagrams.

It would be silly to tell people: *Just be careful.* We have to learn how microscopes and spark chambers, etc. work. It's no easy task. But as we learn, the quality of our observations improves. Learning about the instruments and learning about nature go hand in hand. The same can be said about pictures in mathematics. The fact that many mislead is no reason to reject them in principle as a source of evidence. We simply *have to learn* how to use them, just as we must continue to learn more about microscopes so that we can continue to do better biology. This is a process which will never end.

I realize that Platonistic talk of 'the mind's eye' and 'seeing mathematical entities', is highly metaphorical. This is to be regretted – but not repented. Picture-proofs are obviously too effective to be dismissed and they are potentially too powerful to be ignored. Making sympathetic sense of them is what's required of us.

Calling the Bluff

J.E. Littlewood, a very fine British mathematician, writing at mid-century, remarks: 'My pupils *will* not use pictures, even unofficially and when there is no question of expense. This practice is increasing; I have lately discovered that it has existed for 30 years or more, and also why. A heavy warning used to be given that pictures are not rigorous; this has never had its bluff called and has permanently frightened its victims into playing for safety. Some pictures, of course, are not rigorous, but I should say most are (and I use them whenever possible myself)' (1953/1986: 54).

Littlewood goes on to illustrate with a nice example.

> One of the best pictorial arguments is a proof of the 'fixed point theorem' in one dimension: *Let f(x) be continuous and increasing in $0 \leq x \leq 1$, with values satisfying $0 \leq f(x) \leq 1$, and let $f_2(x) = f\{f(x)\}$, $f_n(x) = f\{f_{n-1}(x)\}$. Then under iteration of f every point is either a fixed point, or else converges to a fixed point.*
>
> For the professional the only proof needed is [Figure 12.6].
>
> (Littlewood 1953/1986: 55)

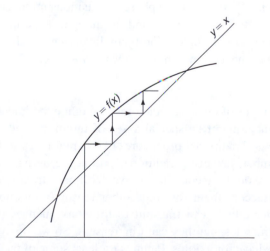

Figure 12.6 Picture proof of a fixed point theorem

This example is completely convincing. Perhaps it should embolden us to greater ambition. There's no reason for not generalizing even more from Littlewood's bluff-calling injunction. He tells us to use regular proofs and picture proofs, as well. But why should that be the end of it? There are good reasons to think we already have several distinct ways of acquiring mathematical knowledge. Here are some of them:

(1) *Proofs*. Traditional verbal/symbolic derivations.

(2) *Intuition*. In the immediate sense that some proposition is true, e.g. $2 + 2 = 4$.

(3) *Induction*. Computing examples, especially if they are the 'right kind' and in sufficient number, can provide evidence of the truth of some mathematical proposition.

(4) *H-D*. I'll use the term hypothetico-deductivism to cover a multitude of variations. But the idea in each case is that we test a mathematical proposition by seeing if its *consequences* are true. The consequences themselves might, for example, be tested by intuition or by induction, or they might already be known to be true. This, presumably, is how axioms that are not directly intuitive (i.e. not self-evident) come to be accepted.

(5) *Pictures*. This can be a prod to intuition, but can also establish very general theorems. In some cases they are just as rigorous as any traditional verbal/symbolic proof.

(6) *Diagonalization*. Gödel proved the existence of an undecidable sentence; but, as a result of the proof, we can see that the sentence is true. This seems unlike the other ways of establishing mathematical results and may be a distinct source of mathematical evidence.

(7) *Thought Experiments*. Some idealized physical situations might correspond to mathematical situations, and a thought experiment involving the former could solve a problem involving the latter. Freiling's darts and the continuum hypothesis are, arguably, an instance.

These are not wholly distinct types of evidence. Pictures, thought experiments, and diagonalization, for instance, might be special cases of intuition. And, of course, they are not all rigorous. Traditional proofs are. Let's grant that for the sake of the argument. Remember, however, traditional proofs are derivations from first principles – but where do these come from? We don't have traditional proofs of the axioms, so we accept them for some other reason – intuition, perhaps, or H-D. Rigorous derivations can transmit to the consequences the confidence we have in first principles, but they can't increase it. So we'll never do better than the evidence we have for axioms. Pictures (at least some of them) stand up well to this comparison. Does anyone really have greater confidence in, say, the axioms of set theory than in the visual proofs I gave in Chapter 3? The same could be said about proofs by diagonalization.

The moral is simply this: there are many different ways to establish mathematical truths. We know a handful; there may be indefinitely many more. Mathematical research since Gödel has been under something of a cloud. The richness of mathematics has been seen as grounds for pessimism – at least in some quarters where it may be argued that even arithmetic is so rich that we cannot hope to capture it. But I see things quite the other way: mathematics is so rich that it offers us techniques for solving *any* problem. In his famous 'On the Infinite', Hilbert wrote stirringly that 'every mathematical problem is solvable . . . we always hear the cry within us: There is the problem, find the answer; you can find it just by thinking, for there is no *ignorabimus* in mathematics' (1925: 200). If we confine ourselves to finitistic methods, or to derivations in first-order logic, or to any other specific set of allowed verbal/symbolic techniques, we shall likely not succeed. But if we allow ourselves all the resources that mathematics offers us (intuitions, pictures, derivations, and other techniques yet undreamed of) then Hilbert, I'm sure, is absolutely right: we will not remain ignorant – any problem can and will be solved.

Math Wars: A Report from the Front

The expression 'Math Wars' could reasonably be taken in several different ways. One is to refer to the nasty debates over calculus reform, especially in the USA. Both sides despair test results of American students (as compared with non-Americans) and the general lack of interest in or knowledge of mathematics. One side wants to re-emphasize skills and rigour; the other looks to innovative ways of teaching mathematics. Terms like 'elitist' and 'flake' are commonplace in this debate about mathematical education.[6]

Another sense of 'math wars' is connected to the 'science wars', the battle raging over social constructivism. Do scientists make discoveries, or do they 'construct' nature in a way that reflects social factors and serves various non-cognitive interests? The debate was brought to a head when Alan Sokal wrote a hoax paper using the worst postmodern jargon. It was unwittingly accepted by the journal *Social Text*. He later exposed his hoax to an uproar that made front-page news. Mathematics has largely gone unnoticed in these debates, but not completely. David Bloor (who anchors the naturalistic wing of social constructivism, which is far removed from the more nihilistic postmodern wing) has attempted a sociological account of mathematics (Bloor 1991). If constructivism continues to flourish, then it will no doubt make incursions here as well.[7]

I'm not interested in either of these math wars, but I mention them to indicate that mathematics lends itself just as readily as anything else to ideological fracas.

And a fracas is just what we have when it comes to 'experimental mathematics'. This is the sense of 'math wars' I now want to take up. Occasionally I'll adopt the stance of the opinionated columnist, but for the most part in this section I will assume the role of the intrepid reporter bringing back news from the front lines.

A story has to begin somewhere – I'll start with a book review. James Gleick (a science journalist) published a very successful book called *Chaos: Making a New Science* (1987). It described the mathematics of 'chaos' in completely non-technical terms, reproduced lots of pictures, described the personalities of some of the major players. And it did all this in a readable and even romantically entertaining way, becoming a best seller. Working mathematicians had mixed opinions of the book, some were quite uneasy. *Chaos* was reviewed very critically in *The Mathematical Intelligencer* by John Franks (1989), a professional mathematician. *The Mathematical Intelligencer* is an entertaining, non-research journal for professional mathematicians; it is filled with expository articles, history, philosophy, biography, reviews, gossip and opinion. It is widely read and quite influential.

Franks finds much to praise in Gleick's work, but on a few essential points he is unsparing: 'One could read this book and come away with the view that mathematical proofs are an obstacle to the pursuit of truth – a sort of self-imposed mental straitjacket worn by stodgy old pedants' (1989: 68).

Alas, Franks goes on to say 'Mathematics has a methodology unique among all the sciences. It is the only discipline in which deductive logic is the sole arbiter of truth' (*ibid.*). I hope that no reader who has read this far gives that claim any credence whatsoever. Franks continues, 'I would contend that an important criterion for judging a scientific discipline is the half-life of its truths' (*ibid.*). Really? Big-bang astronomy hasn't been around for that long, and is currently undergoing 'inflationary' modifications. Before that it was the expanding, steady-state theory, and before that. . . . In short, the beliefs in this realm are not very stable. The accepted truths of astrology, by contrast, haven't changed in aeons; it prides itself on being the repository of ancient wisdom.

Franks likens the hype surrounding chaos and non-linear dynamics to the hype surrounding catastrophe theory a decade earlier – in both cases, he says, it is quite unwarranted. Much of the problem stems from questionable applications, according to Franks. Models must explain. 'It is not enough to find a mathematical system that exhibits similar behavior to a physical experiment, but has no apparent connection with it' (*ibid.*). (This seems a most unlikely doctrine if taken at face value. The main argument of my Chapter 4 on applied mathematics was to the effect that mathematical models explain nothing in the non-mathematical realm. Franks may actually only be complaining that modelling is rather slipshod. He would be right in this – at least much of the time.)

Another prominent mathematician, Morris Hirsch, followed Franks in a later issue of the *Intelligencer* with an opinion piece which praised Gleick's book, but found 'one great defect: It doesn't do justice to the rigorous mathematics underlying a great deal of research in nonlinear dynamical systems' (Hirsch

1989: 6). Whereas Franks thinks that Gleick 'has greatly overestimated the achievements of chaos theorists' (Franks 1989: 69), Hirsch thinks that Gleick has failed to appreciate the rigorous character of their work:

> [T]he earliest and most influential examples [of chaos] were first identified and explored *not* by computer simulation, and *not* by physical experiment, but by *mathematical proof* (Poincaré, Birkoff, Levinson, Smale, Anosov, Kolmogorov, Arnold, Moser . . .). These and many other mathematicians achieved by rigorous mathematical analysis crucial insights into what is now called chaos. It is difficult to imagine that what they discovered could have been found through any kind of experimentation any more than the existence of irrational numbers – which was even more astonishing when it occurred – could have been discovered by computation.
>
> (Hirsch 1989: 6)

Gleick replied testily:

> [T]his seems a bitter pill for some of you [mathematicians] to swallow – there are times when mathematical proof (essential though it is!) comes, historically, as an afterthought. Lorenz's work had its greatest influence before anyone even could say with certainty that his attractor was chaotic. Langford's proof of Feigenbaum was ingenious and admirable, but it did little, really, to validate Feigenbaum's breakthrough – experiments accomplished that. Those mathematicians who choose to look *only* at the documented genealogy of published proofs do their discipline a disservice, I think. It's no wonder they find it awkward or unpleasant to assess Benoit Mandelbrot's place: here is a nominal mathematician who, ostentatiously not proving much, has concretely changed the working lives of many thousands of scientists. You [mathematicians] should not doubt Mandelbrot's powerful originality and importance in the science of our time – though I know some of you still do.
>
> (Gleick 1989b: 9)

Speaking of Mandelbrot (who coined the term 'fractal' and is by far and away the leading public figure in the field), he, too, jumped into the fray. Dismissing Franks's critical remarks on Gleick (of whose book he much approves), Mandelbrot makes a useful distinction between 'top-down' and 'bottom-up' mathematics. The Bourbaki approach is paradigmatic of the top-down method. Start with precise definitions and axioms and then proceed step by rigorous step from there. But Mandelbrot sees himself as working in a messy, open-ended, bottom-up way.[8] Lakatos would be sympathetic. Mandelbrot's hero – 'God on Earth' – is Poincaré, of whom Hermite often complained that he couldn't finish his proofs.

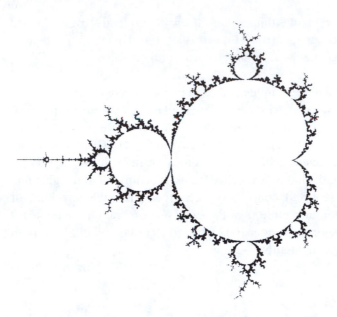

Figure 12.7 The Mandelbrot Set, the set of points in the complex plane that result from the reiteration of $z = z^2 + c$ (where c is a complex number)

Hot on the heals of the Gleick-Franks-Hirsch-Mandelbrot exchange came another in the same journal, this time between Steven Krantz, a prominent analyst, and Mandelbrot. It started with Krantz reviewing two books on fractals for the *Bulletin of the American Mathematical Society*. The review was initially accepted for publication, but then later rejected as too controversial. However, the *Mathematical Intelligencer*, which 'welcomes controversy', gladly published it along with Mandelbrot's reply.

Krantz criticizes much. First, there is no accepted definition of 'fractal'. He mockingly adds, 'It seems that if one does not prove theorems (as, evidently, fractal geometers do not), then one does not need definitions' (1989: 13–14). Second, Fractals have sometimes been called an intellectual advance comparable with calculus. Krantz scoffs at this: 'One notable difference between fractal geometry and calculus is that fractal geometry has not solved any problems' (*ibid.*). Third, the most that fractal geometers can come up with is vague and suggestive ideas. 'The trouble with any subject that relies on more computer output than on theory is that one has to think of something to say about it. The result is that much of the writing turns out to be anecdotal' (*ibid.*). Fourth, the fractal industry is plagued with excessive hype. 'I do not accept the assertion that the Mandelbrot set [Figure 12.7] "is considered to be the most complex object mathematics has ever seen." This type of hyperbole may appeal to readers of popular magazines but rings untrue to the trained mathematician' (*ibid.*: 15). Finally, Krantz allows that computer-generated pictures can be a wonderful tool for investigating mathematical questions. But something has gone wrong when the pictures become an end in themselves. 'In fractal geometry one uses some mathematics to generate a picture, then asks questions about the picture – which generates more pictures. Then one asks more questions about the new pictures. And so on. One rarely, if ever, sees a return to the original mathematics' (*ibid.*: 16). In sum, Krantz thinks fractal geometry is not a new science, nor likely to provide the tools for a new analysis of nature. In short, as he puts it, 'the emperor has no clothes' (*ibid.*).

Mandelbrot replied to this, but the focus of his remarks concerns priority. Krantz had suggested that such things as the 'Mandelbrot set' and the 'Mandelbrot function' were actually developed earlier by others (Julia and Weierstrass, respectively). Needless to say, Mandelbrot took umbrage. But the main point – the legitimacy of this type of mathematics – went largely undefended.

Two distinct issues arise in the field of non-linear dynamics, fractals and chaos: First, what are the standards of rigour in this field, and what should they be? Second, how do we know that a particular application to nature is a legitimate one? Concerning the first point, fractals, etc. seem on fairly solid ground. It's true that some key concepts (including 'fractal' itself) have gone undefined, but that does not separate it from other examples of wonderful mathematical theories, especially in their early days. Riemann, for example, didn't define 'manifold' and a proper definition was rather slow in coming, yet no one would say his work was seriously unrigorous. The crucial idea behind fractals and chaos – iterated mappings – is not in the least problematic. Commentators such as Hirsch are quite right to insist that the highest standards of rigour have been met.

It is the second point, applications, where eyebrows should be raised. In any fractals presentation, claims are made – often quite wild – that such and such a natural phenomena is a fractal or is chaotic. Computer-generated pictures are sometimes highly suggestive of, for example, landscapes, or plants, or coastlines, etc. (See Figure 12.8, a fern, and Figure 12.9, the first few iterations of the

Figure 12.8 A fractal fern

Figure 12.9 The first five iterations of the Koch snowflake

Koch snowflake, a model for a coastline.) But aside from a superficial appearance, there is no evidence or argument that the phenomenon in question is rightly modelled by a fractal.

A favourite example is the coastline of Britain. How long is it? It turns out that we first have to choose a scale. We might start out ignoring inlets and bays; but, at a smaller scale of measure, we would include these, and at a still smaller scale we would include tiny little wiggles. The finer our measuring, the longer the coast. This is how it is with fractals. But is the British coast a fractal? Certainly not. There's a smallest scale – at the subatomic level. And when we measure the boundary particles, quark to quark to quark[9] around the edge, we end up with the longest – but still finite – length. This is not a fractal curve.

The situation here is not unlike the situation a few years ago with Catastrophe Theory. The pure mathematics of Catastrophe Theory is very solid. The discredited hype concerned applications to animal aggression, to games of strategy, to international politics, and so on. In both cases people would do well to reflect on some of the elementary considerations (reviewed in Chapter 4) of how mathematics gets applied to the world.

Two popular, but quite influential, general science magazines got into the act and further fanned the flames of the math wars. William Bowen discussed 'New-wave Mathematics' in *The New Scientist* (1991) and John Horgan announced 'The Death of Proof' in *Scientific American* (1993). Both were subsequently deluged with outraged letters. Each cited a number of the main players, including: Krantz, who is quoted by Bowman as calling research in fractals 'easy, flashy, and . . . pointless'; Mandelbrot, who, of course, defends his own work; and David Epstein, who 'wants experiments to come out of the closet'. Epstein is the editor of a new journal, *Experimental Mathematics*, sometimes derisively called 'The Journal of Unproved Theorems'. Among the most prominent mathematicians to approve of experimental work are John Milnor and William Thurston, both Fields Medal[10] winners for work which would pass traditional muster. Needless to say, they speak with great authority. And, finally, in making their cases for experimentation, both Bowman and Horgan prominently cite the now famous video 'Not Knot' which gives a visual proof of a famous result by Thurston.[11]

The inauguration (in 1992) of the journal *Experimental Mathematics* marks something of a watershed. The first issue included a brief manifesto in which the editors quote Gauss as saying that he discovered mathematical truths 'through systematic experimentation'. But they also declare that they will 'not depart from the established view that a result can only become part of mathematical knowledge once it is supported by a logical proof' (Epstein *et al.* 1992: 1), thereby pleasing everyone – or no one. Their attitude is a bit like George Polya's. He has done as much as anyone to promote heuristic reasoning (with pictures, induction, analogy, etc.) Nevertheless, says Polya, 'We secure our mathematical knowledge by *demonstrative reasoning*. Demonstrative reasoning has rigid standards, codified and clarified by logic (formal or demonstrative logic), which is the theory of demonstrative reasoning' (1954: v). But it is not the stated aims of *Experimental Mathematics* which are so important as the mere presence of the journal. Its very existence encourages and legitimizes 'experimental' mathematics.[12]

By far and away, the most important and influential debate in the math wars is that which took place in the *Bulletin of the American Mathematical Society*. Arthur Jaffe and Frank Quinn began with their article on 'theoretical' vs. 'rigorous' mathematics (1993). They start by noting that rigour is double-edged. It has 'brought to mathematics a clarity and reliability unmatched by any other science. But it also makes mathematics slow and difficult' (*ibid.*: 1). They offer the following picture of mathematical work, a picture which would be widely accepted by working mathematicians: 'Typically, information about mathematical structures is achieved in two stages. First, intuitive insights are developed, conjectures are made, and speculative outlines of justifications are suggested. Then the conjectures and speculations are corrected; they are made reliable by proving them' (*ibid.*). They use the terms 'theoretical' for the first stage and 'rigorous' for the second.

'Rigorous' evidently means properly derived via correct logical principles. But derived from what? The axioms or first principles that are used in any proof are themselves not proven, so what is their status in the rigorous vs. theoretical taxonomy? They would have to count as theoretical, but anything that rests on theoretical premises will have to be considered theoretical itself, so it is difficult to recognize that anything could count as rigorous by their lights. Of course, we might take anything derived from axioms as really saying: If the *Axioms* are true then so is the *Theorem*. But this just trivializes mathematical activity into establishing logical truths. Making a worthwhile distinction between 'theoretical' and 'rigorous' is going to be difficult to say the least.

Moreover, purportedly rigorous proofs are not transparently obvious. The recent episode of Andrew Wiles's proof of Fermat's Last Theorem shows that dramatically. As I mentioned earlier, his first go at it was, after initial acceptance, considered by those in the field to be incomplete and flawed. After a great deal of extra work the revised proof is now thought to be correct; but how much faith should anyone put in it?

Logical proofs do two things. One of these is epistemic: they provide evidence that a proposition is true. They may do this well or poorly. (We should always keep in mind that when proofs serve as evidence, it is *evidence for us*.) The other thing that proofs do is establish connections between propositions, and they can be greatly cherished for this. So, even when we have no particular reason to believe some set of axioms, we might still try to establish connections among them and other propositions, simply to see how the whole structure hangs together. This means that even where we have perfectly rigorous proofs of all theorems, there are still no grounds for believing a single one. The sharp dichotomy with speculation or conjecture is simply false. Of course, a dichotomy is possible, but it is not simple and straightforward.

Jaffe and Quinn cite many cases where speculations and conjectures have been highly stimulating and have benefited mathematics greatly. But they also point to several where things have not gone so well. And with the growing influence of 'theoretical' work, they fear the worst. What are the problems? As they see it:

(1)   Theoretical work, if taken too far, goes astray because it lacks the feedback and corrections provided by rigorous proof.

(2)   Further work is discouraged and confused by uncertainty about which parts are reliable.

(3)   A dead area is often created when full credit is claimed by vigorous theorisers: there is little incentive for cleaning up the debris that blocks further progress.

(4)   Students and young researchers are misled.

(Jaffe and Quinn 1993: 8)

Jaffe and Quinn suggest the following prescription for what ails experimental mathematics. They want 'theoretical' work explicitly called such, and when someone later finds a rigorous proof, the credit for the result must be appropriately shared. They want terms like 'conjecture' used instead of 'theorem' and 'supporting argument' instead of 'proof'. Ideally, titles would include 'theoretical' or 'speculative' so that readers clearly understand their status. It is, they suggest, 'mathematically unethical not to maintain the distinctions between casual reasoning and proof' (1993: 12).

Jaffe and Quinn seem like the sweet voice of reason, perhaps even good old-fashioned mathematical common sense. After all, everyone distinguishes between the Pythagorean *theorem* and the Riemann *hypothesis*, between the Prime Number *theorem* and Goldbach's *conjecture*. How could anyone object?

Yet object they did. In a subsequent issue of the *Bulletin* a large number of mathematicians, including several very prominent Fields Medal winners, took on Jaffe and Quinn and their theoretical/rigorous distinction (Atiyah *et al.* 1994). Michael Atiyah complained that Jaffe and Quinn 'present a sanitized view of mathematics which condemns the subject to an arthritic old age' (Atiyah *et al.* 1994: 178). René Thom linked 'rigour' with 'rigor mortis' (*ibid.*: 204). Armand Borel objected that 'what mathematics needs least are pundits who issue prescriptions or guideline for presumably less enlightened mortals' (*ibid*.: 180). Mandelbrot regretted that Jaffe and Quinn had taken up 'tribal and territorial issues' and found what they had to say 'appalling' (*ibid*.: 193). Karen Uhlenbeck accused them of a 'too narrow perspective' (*ibid*.: 201), as did Edward Witten who believed they had a 'rather limited idea' of how mathematics and physics interact (*ibid*.: 205).

In a separate and much longer article, but still part of the rejoinders to Jaffe and Quinn, William Thurston objected to the 'one-dimensional scale (speculation vs. rigour)' (Thurston 1994: 161), and raised a number of interesting points while defending the more 'experimental' activities of himself and others.

We could characterize mathematicians as theorem provers and ask how they do it. But Thurston thinks this is the wrong way to go about things. Instead, mathematicians should be seen as people who 'advance human understanding of mathematics' (*ibid*.: 162), and the interesting question is: How do they do that? Thurston's emphasis throughout is on *human understanding*, 'which does not work on a single track, like a computer with a single central processing unit' (*ibid*.: 164). Proof seems central, but what is it? Thurston notes the usual problems with incomplete or downright false proofs. But he thinks that recorded, correct proofs are not all that important, because 'Mathematical knowledge and understanding were embedded in the minds and in the social fabric of the community of people thinking about a particular topic. This knowledge was supported by written documents, but the written documents were not really primary' (*ibid*.: 169). The point is that people in the field have a grip on the structure, and through conversation, diagrams, a few coded expressions, etc. can quickly convince others in the field of the correctness of any particular claim. Published, correct proofs mean little to the life of any particular mathematical field according to Thurston.

Who are these people objecting to the Jaffe–Quinn view which is an expression of orthodoxy? William Thurston, Fields Medal 1982, Director of Mathematical Sciences Research Institute, Berkeley, is a leading geometer. Edward Witten, Fields Medal 1990, a champion of string theory in physics, is often said to be the most brilliant physicist working today. René Thom, Fields Medal 1958, is the creator of 'catastrophe theory'. Michael Atiyah, Fields Medal 1966, Master of Trinity College and Director of the Mathematical Institute at Cambridge, is one of the most prominent and influential mathematicians of recent times. I mention who they are to show something important about the debate. The 'rebels' are not young turks from the fringe; they are pillars of the mathematical community. I'll turn now to offering a few brief remarks on some details of the debate in the math wars.

One false assumption running throughout these discussions is this: if mathematics is done more like physics, then there will be a loss of certainty. For some aspects of experimental mathematics this will be true. But not all, and it's important to get a better idea of how physics sometimes works. Thought-experiments offer a clear counter-example to this claim.

Let's look at one of the finest examples: Galileo's wonderful argument in the *Discoursi* to show that all bodies, regardless of their weight, fall at the same speed. It begins by noting Aristotle's view that heavier bodies fall faster than light ones (H > L). We are then asked to imagine that a heavy cannon ball is attached to a light musket ball. What would happen if they were released together? (Figure 12.10).

Reasoning in the Aristotelian manner leads to an absurd conclusion. First, the light ball will slow up the heavy one (acting as a kind of drag), so the speed of the combined system would be slower than the speed of the heavy ball falling alone (H > H + L). On the other hand, the combined system is heavier than the heavy ball alone, so it should fall faster (H + L > H). We now have the absurd consequence that the heavy ball is both faster and slower than the even heavier combined system. Thus, the Aristotelian theory of falling bodies is destroyed.

But the question remains, 'Which falls faster?' The right answer is now plain as day. The paradox is resolved by making them equal; they all fall at the same speed (H = L = H + L).[13]

Much science is based on inductive leaps from a limited body of data. But not all. Galileo's thought-experiment is just as rigorous as any proof found in any mathematical journal. We have little to fear if mathematical reasoning becomes like that supreme example of human ingenuity in the natural sciences.

The question of just what counts as rigour is actually not settled in the debates I have been describing. Back in Chapter 3 I claimed some of the picture-proofs found there are rigorous. If this is correct (though I suspect Jaffe and Quinn would only count logical derivations as rigorous), then there is

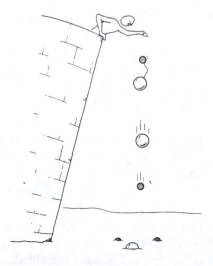

Figure 12.10 Galileo's thought experiment establishing law of free-fall

considerable scope for compromise. Much experimental mathematics is visual. As such it is often dismissed as merely 'heuristic.' It may well be both heuristic and rigorous. The trade-off – assumed to some extent by all in this exchange – between rigour and understanding is simple-minded.

Once More: The Mathematical Image

In the opening chapter I listed a number of ingredients of the mathematical image, features that the vast majority of working mathematicians, philosophers, and ordinary people would endorse as characteristic of mathematics (reproduced below in italics). How have these points stood up? A word or two in connection with each should suffice.

(1) *Mathematical results are certain*
(2) *The history of mathematics is cumulative*

Not a hope for either of these. Mathematics is certainly fallible; no theorem is immune from overthrow. The history of mathematics is filled with revisions, most often in the form of conceptual change, redefinition of key terms.

(3) *Mathematics is objective*

Happily accepted, and the objectivity is as Platonism characterizes it.

(4) *Proofs are essential*

Not if proof means logical derivation from first principles. The scope of mathematical evidence is very much wider than that. There are picture-proofs and more.

(5) *Diagrams are psychologically useful, but prove nothing*
(6) *Diagrams can even be misleading*

Diagrams can certainly mislead; the trick is to find ways of anticipating this. In any case, pictures are often very much more than psychological aids – they can be rigorous proofs.

(7) *Mathematics is wedded to classical logic*

'One god, one flag, one logic', said Whitehead. Well, he was at least right about logic, and it is classical.

(8) *Mathematics is independent of sense experience*

Right again. This is the sense in which mathematics is a priori. It is not infallible, however. Experimental mathematics, it must be stressed, should not be confused with mathematics based on sensory evidence – there's no such thing.

(9) *Computer proofs are merely long and complicated regular proofs*

This is probably more of an open question than the others. But the point is likely right, if we take computers to be tools that extend our mathematical ability, as telescopes extend our vision. Both, of course, are fallible.

(10) *Some problems are unsolvable in principle*

Though the issue is vague, I'm convinced that this principle is false. The reason for thinking there are unsolvable problems – a moral often drawn in the light of various undecidability results – is based on assuming specific tools for problem solving, such as first-order logic. But if we allow that there are tools for problem solving not yet discovered, then all pessimism is undermined. I have no grounds for thinking some clever picture will prove the Reimann hypothesis – but I have no reason to deny it, either. Optimism about this or that has often been unjustified, but optimism never gets in the way of progress. Let's take the leap with Hilbert: In mathematics there is no *ignoramibus*.

I'll belabour things only a moment more. Readers convinced of the case made here for Platonism will uphold the objectivity of mathematics, the link to classical logic, and its independence from sense experience. On the other hand, Platonism (as I've characterized it, especially in Chapter 2) means that mathematics is quite fallible, so certainty and a cumulative history of mathematics must be abandoned. Because of this (at least in part) it is easy to accept the fallibilism of very long computations and still say they are nevertheless regular proofs. The biggest departure, no doubt, from the standard picture of mathematics pertains to pictures and thought experiments. Of course, they can be misleading. But they can provide solid evidence, too, evidence which is as rigorous as any traditional verbal/symbolic proof.

Mathematics is one of humanity's nobler activities. Trying to make sense of it is an enormously difficult undertaking, yet filled with the potential of great rewards. Traditional issues are still with us and still important. Platonism, formalism, constructivism and so on have all contributed much to the understanding of mathematics and will surely contribute more. But the real action today – the living philosophical issues for working mathematicians – cluster around visualization and experimentation. The clarification of this cluster of problems presents us with our greatest challenge. If our philosophical duty lies anywhere it is here, after the elimination of war and poverty which, of course, must be the first duty of all.

Further Reading

Tymoczko (ed.), *New Directions in the Philosophy of Mathematics* contains several papers recommending new and diverse approaches to mathematics. In general there is little written on this that is at all systematic, but given its 'newness', we shouldn't expect it. Aside from chatting with working mathematicians (who are often happy to gossip about trends), the best place to find things is in the Letters sections and among the more informal submissions to professional journals such as *Bulletin of the American Mathematical Society*, *Mathematical Intelligencer*, or in Internet discussion groups who often have interesting debates.

Notes

Chapter 2

1 For a study of the historical Plato and Platonism, see Moravcsik (1992). Plato's *Meno* and *Republic* (especially Book IV) are the best introduction to the mathematical aspects of Plato's thought.

2 See Hale (1987) for a very thorough discussion of this issue. I should note that some like Hale and Wright call themselves Platonists, but would not accept all the principles listed here.

3 Not all realists are sanguine about this point. Maddy and some structuralists (Resnik and Shapiro) have considerable sympathy with Platonism, but would qualify the claim that *all* mathematical objects are outside space and time. Their respective views will be taken up in later chapters.

4 There are, however, some echos of the earlier notion in some definitions of number. Russell, for example, defined two as the set of all sets having two members. This suggests a universal.

5 Alexander Bird points out that this needs fine tuning. Games like chess are abstract entities, yet have a history (rule changes, etc.), so that would seem to put them in time.

6 See, for example, Chihara (1973) and Benacerraf (1973). Maddy (1990) takes the problem seriously and looks for a naturalistic way around it which is still compatible with realism.

7 This sort of argument goes back at least to Sextus Empiricus; current versions stem from Benacerraf (1973); others include Kitcher (1983) and Field (1980).

8 For an elementary exposition of EPR and the Bell results see Albert (1992).

9 This example is developed in a bit more detail in Brown (1990, 1991).

10 Kitcher's other arguments against Platonism are of the no-access sort dealt with above.

11 A spectrum of views can be found in the articles by Boolos, Parsons, and Wang in Benacerraf and Putnam (1983).

12 See, for example, Brown (1991, 1994) for more on laws of nature and thought experiments.

Chapter 3

1 Some other recent authors have taken a positive view of pictures. For example, see Barwise and Etchemendy (1991), Shin (1994), Giaquinto (1994) and Hammer (1995).
2 Bolzano uses concepts like *least upper bound* and *greatest lower bound* which he employs in the following theorem: If a property M does not hold of all values of a variable quantity x but holds of all those which are less than a certain quantity u, then there is always a quantity U, which is the largest of those quantities y which are such that every $x < y$ has property M.
3 Years later in his autobiographical sketch he remarked that he had 'found a way to derive from concepts many geometric truths that were known before only on the basis of mere visual appearance' (1969–87: vol. 12, 68).
4 Giaquinto (1994) raises objections based on examples of continuous but nowhere differentiable functions, since they cannot be visualized. This is an important point. However, I'm not persuaded of its seriousness, since I see no need to visualize each and every continuous function in order to justify the theorem. After seeing a single example, we can grasp the general point. It holds for smooth and non-smooth, algebraic and transcendental, and so on.
5 See Gary and Johnson (1978) for the standard, yet still accessible, work on this topic.
6 For more on Russell see Irvine (1989).
7 Excellent discussions can be found in Aleksandrov (1963) and Friedman (1983).
8 A great deal more needs to be said about this. For one thing, someone might say '9 has the property of being the number of the planets' and then note that this is hardly an essential property of 9, so numbers have accidental properties, too. Perhaps we can get around this by distinguishing between *intrinsic* and *extrinsic* properties. My weight is an intrinsic property of me. Suppose I have the property of having the same weight as Bob. If he goes on a diet I lose that property; but in a perfectly reasonable sense I have not changed, and that's because an extrinsic property has changed, not an intrinsic one. So, being the number of the planets, we could say, is an extrinsic property of 9, but being a composite number is intrinsic. If something like this works, then take what I said about all mathematical properties being essential as being a claim about intrinsic mathematical properties. I should admit that I'm not at all sanguine about this point or related ones made earlier. I'm grateful to Alexander Bird, Bill Newton-Smith and Mary Tiles for critical discussions.

Chapter 4

1 For more on laws of nature see Brown (1994: chs 6, 7 and 8).
2 To see this phenomenon at work in a particularly interesting and controversial way, consider the interaction between quantum field theory and knot theory in the work of Witten (1989).
3 See Devlin *et al.* (1997) for a good collection of articles which picks *The Bell Curve* apart and makes its pseudo-scientific status evident.
4 The Koblitz and Simon series of articles are in the 1988 volume of *The Mathematical Intelligencer*. Further references can be found there.
5 See Papineau (1993) for a 'fictionalist' variant of this view.
6 The type of cause would not be efficient, to use Aristotle's terminology, but rather would be formal.
7 The recent books, Resnik (1997) and Shapiro (1997) appeared too late for me to do justice to them. The main features of their views seems unchanged from their earlier accounts.

Chapter 5

1 See Reid (1970) for a biographical account.
2 For a history of the infinite see A.W. Moore (1990). For an account of set theory which is more constructive in spirit and less sympathetic to actual infinities, see Tiles (1989).
3 See Coxeter (1974) or Courant and Robbins (1941).
4 (a)

	(1)	$\forall x(\sim(0 = sx))$	Axiom (2)
	(2)	$\sim(a = sa)$	Assumption
	(3)	$\forall x \forall y(sx = sy \rightarrow x = y)$	Axiom (1)
	(4)	$\forall y(sa = ssy \rightarrow a = sy)$	(3) Universal instantiation
	(5)	$(sa = ssa \rightarrow a = sa)$	(4) Universal instantiation
	(6)	$\sim(sa = ssa)$	(2), (5) Modus tollens
	(7)	$\sim(a = sa) \rightarrow \sim(sa = ssa)$	(2), (6) Conditional proof
	(8)	$\forall x(\sim(x = sx) \rightarrow \sim(sx = ssx))$	(7) Universal generalization
	(9)	$(\forall x(\sim(0 = sx))$ & $\forall x(\sim(x = sx) \rightarrow$ $\sim(sx = ssx))) \rightarrow \forall x(\sim(x = sx))$	Axiom (8)
	(10)	$\forall x(\sim(x = sx))$	(1), (8), (9) Modus ponens

5 Some typical texts with good expositions are: Mendelson (1987) or Boolos and Jeffrey (1989). The second of these presents Gödel's theorem via Turing machines.
6 Suppose $F(x)$ names both n and m. Thus, we have $\forall x(F(x) \rightarrow x = [n])$ and $\forall x(F(x) \rightarrow x = [m])$. Therefore, $\forall x(F(x) \rightarrow (x = [n]$ & $x = [m]))$; hence $[n] = [m]$; and so, $n = m$.
7 There are only 16 primitive symbols in the formal system, so there can only be at most 16^i formulae containing i symbols. Hence there can only be at most 16^i numbers named by formulae containing i symbols. And this, of course, is a finite number.
8 Recall that [10] is the expression ssssssssss0 which has 11 symbols. In general, $[n]$ will be the expression containing n successor symbols and a 0; hence, $[n]$ will contain $n + 1$ symbols.
9 Suppose $\vdash (p \rightarrow q)$, then $\vdash \exists \varphi(p \rightarrow q)$ follows by Rule I; then $\vdash \exists \varphi(p \rightarrow q) \rightarrow (\exists \varphi p \rightarrow \exists \varphi q)$ follows by Rule II; finally we get $\exists \varphi p \rightarrow \exists \varphi q$ by *modus ponens*. And so we have our result, if $\vdash (p \rightarrow q)$ then $\vdash (\exists \varphi p \rightarrow \exists \varphi q)$.
10 See Vesley (forthcoming).

Chapter 6

1 Chemistry is a serious contender. We can tell a lot about a molecule, say, water or ethanol, just by looking at its chemical name: H_2O or C_2H_5OH, respectively.
2 More accurately and more generally, it is 'base n' notation that is clever rather than the Arabic numerals specifically.
3 Since each crossing has two numbers, there must be in total an even number of odd and even labels. If we start with any crossing labelled n, then move along the curve, we must cross strands an even number of times, say $2k$, in order to get back to the initial crossing, since each time we go 'out' we must cross back to get 'in'. The second label at our starting crossing, m, must be of the form $n + 2k + 1$. If n is odd, then m is even, and vice versa.

Chapter 7

1 See, for example, Suppes (1957) for an extensive discussion.
2 See Demopoulos (1995). This is a collection of important recent papers which focus on what did and what did not work in Frege's original logicism. See also Apostoli (forthcoming).
3 For an introduction to graph theory see Trudeau (1976). A good, advanced work is Bollobás (1979).
4 I'm especially grateful to my Toronto colleague Alasdair Urquhart for pointing out several interesting features of graph theory.

Chapter 8

1 For a history of the infinite which is sympathetic to constructivist views, see Moore (1990).
2 For a discussion of the technical results and their consequences for constructivism, see Hellman (1993a, 1993b, 1993c).
3 Indeed, current computer algorithms used in cryptanalysis for establishing primality do not generally yield explicit factors for composite numbers.

Chapter 9

1 It is hard to find an example which makes the student look anything but silly. In this case (as Mary Tiles reminds me) the properties of squares of numbers (i.e. n^2 is bigger than n) seem to be violated. This could be used to justify saying that the student is wrong.
2 An important qualification, stemming from quantum mechanics is needed. A radio-active atom has a disposition to decay. But underlying categorical facts associated with this would amount to hidden variables, something which is greeted with much scepticism in many quarters. By analogy, the mind might have dispositions without underlying categorical facts. Needless to say, this is all quite controversial.
3 See for example Goldfarb (1985), Collins (1992), McDowell (1992), and many others.

Chapter 10

1 This was the result of several hundred hours on an IBM 370-160A, a state-of-the-art computer in 1976.
2 For those who want the definition, here it is: A *finite projective plane of order n* (where $n > 0$) is a set of $n^2 + n + 1$ lines and the same number of points, such that: (1) every line contains $n + 1$ points, (2) every point lies on $n + 1$ lines, (3) any two distinct lines intersect at exactly one point, and (4) any two distinct points lie on exactly one line. The seven-point geometry is a projective plane of order 2 because it meets these conditions, i.e. it has $2^2 + 2 + 1 = 7$ lines, etc.

3 Leading word-processing programmes, for example, will have thousands of so-called beta testers.

4 This clearly undermines Paul Teller's point in his argument against Tymoczko: 'If one repeats a proof of a fact about numbers, unlike a measurement of the charge of the electron, one has to get the same result as before, again on the assumption that one does not use a mistaken method of proof and as long as one makes no mistake in applying the method of proof' (Teller 1980: 802). In the case of the CRAY, the same hardware running the same software, with the same input can give different results on different occasions. Perhaps Teller would reply that interference by a cosmic ray meant that the assumption of no mistake in applying the method of proof was violated. But if he adopts this position, it is hard to see how anyone could, for example, ever get different values for the measured charge of an electron – on the assumption that no mistakes in measuring were made.

5 See Kari (1997) or Fallis (1996) for expositions, and Fallis (1997) for a philosophical discussion.

6 A *group* is a set of objects with operations defined on them. It's *finite* when there are only a finite number of elements in the group, and it's *simple* when it doesn't have normal subgroups. Isomorphic groups are taken to be just a single group. A good elementary discussion can be found in Davis and Hersh (1981).

7 Thanks to powerful computers and intense interest, this number has become very unstable. A short while ago (1995) only 31 were known. While writing successive drafts of this chapter, I've had to change this number three times. It's sure to be out of date soon. (That's what I said in the first edition. Now (2007) the number is up to 45.)

8 If you have a PC, access to the Internet, and a taste for such things, you can join 'The Great Internet Mersenne Prime Search' at http://ourworld.compuserv.com/homepages/justforfun/prime.htm

9 For those who have forgotten, logarithms are characterized as follows: $log_a b = c$ if and only if $a^c = b$. Thus, for example, $log_2 5 = 2.32 \ldots$ because $2^{2.32 \cdots} = 5$. In Figure 10.1 the numbers on the horizontal axis represent the nth Mersenne prime. The p in log_2 is the prime in $2^p - 1$, so $p = 7$, and $log_2 7 = 2.8$.

10 The example comes from Dunham (1994: 117).

11 The commonly used foumula currently is due to Chudnovsky and Chudnovsky,

$$\frac{1}{\pi}\frac{640320^{3/2}}{6541681608} = \sum_{n=0}^{\infty}\left(n + \frac{13591409}{545140134}\right)\frac{(6n)!}{(3n)!(n!)^3}\frac{(-1)^n}{640320^{3n}}$$

12 See various articles by J. Borwein, P. Borwein *et al.* for more examples of this sort and for discussions of relevant statistical analyses of computer-generated data.

Chapter 11

1 Proper classes contain members, just as sets do, but they are not themselves members of other sets. They are sometimes said to be 'too big' to be sets. If we tried to form a set of all ordinals, it would have led to a contradiction, so such a set cannot exist. The proper class of all ordinals does not lead to this absurdity.

2 Mathematical usage differs from ordinary usage, where 27th would be called the ordinal. But the underlying idea is the same, since ordinal signifies order in both cases.

3 Gödel's views are found in his work (1948/64). For a discussion of recent research on CH, see Woodin (2001).

4 These quotes are from the FOM discussions on Freiling over several days in August and September, 2004. They can be found in the FOM archive which is located at http://www.cs.nyu.edu/pipermail/fom/

5 I will amplify a bit by proposing the following rule: Accept a physical version of a mathematical proposition *unless* it is explicitly known to be misleading. This is just like the common sense principle: Accept the evidence of the senses in any situation *unless* the situation is one in which illusions, etc. are known to happen. The upshot is simple: Freiling's dart example wins by default. The onus is on the nay-sayers to show it to be misleading.

6 By calling it essential, I mean only that ZFC need something in addition to prove ~CH. There are, of course, other possibilities besides the dart thought experiment that will lead to this conclusion.

Chapter 12

1 Nicholas Bourbaki is the fictional character used by a group of outstanding French mathematicians, including: Artin, Cartan, Cartier, Dieudonné, Serre, Weil, and others. Over the years, several texts have been published under the Bourbaki *nom de plume*, always upholding the highest standards of rigour, each devoid of diagrams, and generally agreed to be pedagogically hopeless.

2 The paradoxes were only one reason for the interest in rigour. Perhaps even more important was the status of the Axiom of Choice and Zermelo's proof that every set could be well-ordered. See Moore (1982).

3 This is the same Brouwer of constructivist fame; he was also a brilliant topologist. His later commitment to constructivism made him repudiate some of his earlier results.

4 For more on the wonderful topic of space-filling curves, see Sagan (1994).

5 Figure 12.4(b) conceals an overlap of some of the pieces. That's where the missing area is. To grasp the problem, compute the ratio of height to base of the two triangles, the smaller with base 13 and height 5, the larger with base 21 and height 8. They will be exactly the same for similar triangles, but these aren't. The 'diagonal' in the rectangle isn't a diagonal after all – it's not a straight line.

6 The ongoing debate can be found in almost any journal that serves the general needs and interests of mathematicians, for example *Notices of the American Mathematical Society* or *The Mathematical Intelligencer*.

7 I won't take up the issue of social construction here. For more detail see Brown (1989 and forthcoming).

8 Mandelbrot has often complained about the oppressive Bourbaki influence on French mathematics and reports that he left France to escape it.

9 I'll ignore the difficulties quantum mechanics presents in *locating* elementary particles. Addressing this problem would not help the case for fractals.

10 Fields Medals (usually four) are presented at the International Congress of Mathematics, a huge conference which is held every four years. In the mathematical community Fields Medals are taken to be equivalent in status to a Nobel Prize.

11 University libraries with a video collection might well have a copy of this popular video. A glimpse of it (and many other interesting things as well) can be seen at the website maintained by the Geometry Center at the University of Minnesota: www.geom.umn.edu

12 Surprisingly, given his impatience with non-traditionalists, Steven Krantz co-authored a contribution to the first issue.

13 For more on thought-experiments see Brown (1991).

Bibliography

Abraham, R.H. and Shaw, C.D. (1983) *Dynamics: The Geometry of Behavior*, 3 vols., Santa Cruz, CA: Aerial Press.

Adams, C. (1994) *The Knot Book*, New York: Freeman.

Adleman, L. (1994) 'Molecular Computation of Solutions to Combinatorial Problems', *Science* 266: 1021–24.

Albers, D.J. and Alexanderson G.L. (eds.) (1985) *Mathematical People*, Cambridge, MA: Birkhäuser.

Albert, D. (1992) *Quantum Mechanics and Experience*, Cambridge, MA: Harvard University Press.

Alecksandrov, A.D. (1963) 'Curves and Surfaces', in A.D. Alecksandrov *et al.*, (eds), *Mathematics: Its Content, Method, and Meaning*, Cambridge, MA: MIT Press.

Apostol, T. (1967) *Calculus*, Waltham, MA: Blaisdell.

Apostoli, P. (forthcoming) 'Logic, Truth, and Number: The Elementary Genesis of Arithmetic', in A.C. Anderson and M. Zeleny (eds), *Alonzo Church Memorial Volume*, Dordrecht: Kluwer.

Appel, K. and Haken, W. (1976) 'Every Planer Map is Four Colorable', *Bulletin of the American Mathematical Society* 82: 711–12.

Appel, K., Haken, W. and Koch, J. (1977) 'Every Planer Map is Four Colorable', *Illinois Journal of Mathematics* XXI, 84: 429–567.

Arnheim, R. (1969) *Visual Thinking*, Berkeley, CA: University of California Press.

Atiyah, M. *et al.* (1994) 'Responses to "Theoretical Mathematics": Towards a Cultural Synthesis of Mathematics and Theoretical Physics' by A. Jaffe and F. Quinn', *Bulletin of the American Mathematical Society* 30(2): 178–211.

Ayer, A.J. (1936/1971) *Language, Truth, and Logic*, London: Penguin.

Azzouni, J. (2004) *Deflating Existential Consequence*, Oxford: Oxford University Press.

—— (2006) *Tracking Reason*, Oxford: Oxford University Press.

Balaguer, M. (1998) *Platonism and Anti-Platonism in Mathematics*, Oxford: Oxford University Press.

Barwise, J. and Etchemendy, J. (1991) 'Visual Information and Valid Reasoning', in W. Zimmerman and S. Cunningham (eds), *Visualization in Teaching and Learning Mathematics*, Mathematical Association of America.

Bell, J. and Machover, M. (1977) *Course in Mathematical Logic*, Amsterdam: North Holland.

Bell, John, L. (1985) *Boolean-Valued Models and Independence Proofs in Set Theory*, Oxford: Oxford University Press.

Benacerraf, P. (1965) 'What Numbers Could Not Be,' reprinted in Benacerraf and Putnam (1983).

Benacerraf, P. (1973) 'Mathematical Truth', reprinted in Benacerraf and Putnam (1983).

Benacerraf, P. and Putnam, H. (eds) (1983) *Philosophy of Mathematics*, 2nd edition, Cambridge: Cambridge University Press.

Binmore, K. and Davies, J. (2001) *Calculus Concepts and Methods*, Cambridge: Cambridge University Press.

Bishop, E. and Bridges, D. (1985) *Constructive Analysis*, New York: Springer.

Bloor, D. (1991) *Knowledge and Social Imagery*, 2nd edition, Chicago: University of Chicago Press.

Bollobás, B. (1979) *Graph Theory*, New York: Springer.

Bolzano, B. (1817) 'Rein analytischer Beweis des Lehrsatzes, dass zwischen je zwei Werthen, die X (ein entgegengesetztes Resultat gewhren, wenigstens eine reele Wurzel der Gleichung liege, Gottlieb Hass, Prague l.

—— (1851) *Paradoxes of the Infinite*, trans. by D.A. Steele, London: Routledge and Kegan Paul, 1950.

—— (1969–87) *Gesamtausgabe,* 15 vols., Stuttgart: Fromman.

Boolos, G. (1989) 'New Proof of the Gödel Incompleteness Theorem', *Notices of the American Mathematical Society.*

—— (1994) 'Gödel's Second Incompleteness Theorem Explained in Words of One Syllable,' *Mind.*

—— and Jeffrey, R. (1989) *Computability and Logic*, 3rd edition, Cambridge: Cambridge University Press.

Borwein, J. (1997) 'Brouwer-Heyting Sequences Converge', *Mathematical Intelligencer*

—— *et al.* (1992) 'Some Observations on Computer Assisted Analysis', *Notices of the American Mathematical Society* 39: 825–29.

—— *et. al.* (1996) 'Making Sense of Experimental Mathematics', *Mathematical Intelligencer* 18(4): 12–18.

Bowen, W. (1991) 'New-Wave Mathematics', *New Scientist* (August).

Boyer, C.B. (1949) *The Concepts of the Calculus*, New York: Dover (reprint).

Brouwer, L.E.J. (1913) 'Intuitionism and Formalism', reprinted in Benacerraf and Putnam (1983).

—— (1948) 'Consciousness, Philosophy, and Mathematics', reprinted in Benacerraf and Putnam (1983).

—— (1952) 'Historical Background, Principles and Methods of Intuitionism', *South African Journal of Science.*

Brown, J.R. (1989) *The Rational and the Social*, London: Routledge.

—— (1990) 'π in the Sky' in Irvine (1990).

—— (1991) *The Laboratory of the Mind: Thought Experiments in the Natural Sciences*, London and New York: Routledge.

—— (1994) *Smoke and Mirrors: How Science Reflects Reality*, London and New York: Routledge.

—— (2001) *Who Rules in Science: An Opinionated Guide to the Wars*, Cambridge, MA: Harvard University Press.

Burgess, J. and Rosen, G. (1997) *A Subject with No object*, Oxford: Oxford University Press.

Campbell, N. (1920) *Physics: The Elements*, Cambridge, MA: Harvard University Press (reprinted by Dover as *Foundations of Science*, 1957).

Cantor, G. (1895/1897) *Contributions to the Founding of the Theory of Transfinite Numbers*, trans. P.E.B. Jourdain, New York: Dover, 1955.

Chaitin, G. (1975) 'Randomness and Mathematical Proof', *Scientific American.*

Chihara, C. (1973) *Ontology and the Vicious Circle Principle*, Ithaca, NY: Cornell University Press.

—— (1982) 'A Gödelian Thesis Regarding Mathematical Objects: Do They Exist? And Can We Perceive Them?' *Philosophical Review.*

—— (1990) *Constructability and Mathematical Existence*, Oxford: Oxford University Press.

Chihara, Charles S. (2004) *A Structural Account of Mathematics*, Oxford: Oxford University Press.

Coffa, J.A. (1991) *The Semantic Tradition from Kant to Carnap*, Cambridge: Cambridge University Press.

Cohen, Paul J. (1966) *Set Theory and the Continuum Hypothesis*, Reading, MA: Benjamin.

Cohen, Paul J. and Hersh, Reuben (1967) 'Non-Cantorian set theory', *Scientific American* 217(6): 104–16.

Collins, A. (1992) 'On the Paradox that Kripke finds in Wittgenstein', in P.A. French *et al.* (eds), *Midwest Studies in Philosophy*, vol. XVII, Notre Dame: University of Notre Dame Press.

Colyvan, M. (2001) *The Indispensibility of Mathematics*, Oxford: Oxford University Press.

Corfield, D. (2003) *Towards a Philosophy of Real Mathematics,* Cambridge: Cambridge University Press.

Courant, R. and Robbins, H. (1941) *What is Mathematics?*, Oxford: Oxford University Press.

Courant, R., Robbins, H. and Stewart, I. (1996) *What is Mathematics*? 2nd edition, Oxford: Oxford University Press.

Coxeter, H.S.M. (1974) *Projective Geometry*, 2nd edition, Toronto: University of Toronto Press.

Davis, P.J. and Hersh, R. (1981) *The Mathematical Experience*, Boston: Birkhauser.

Dedekind, R. (1888) 'The Nature and Meaning of Numbers', trans. in *Essays on the Theory of Numbers*, New York: Dover, 1963.

Demopoulos, W. (ed.) (1995) *Frege's Philosophy of Mathematics*, Cambridge, MA: Harvard University Press.

Detlefsen, M. (1986) *Hilbert's Programme*, Dordrecht: Reidel.

—— (1990) 'Brouwerian Intuitionism', reprinted in Detlefsen (1992b).

—— (1992a) 'On an Alleged Refutation of Hilbert's Program Using Gödel's First Incompleteness Theorem', in Detlefsen (1992b).

—— (ed.) (1992b) *Proof, Logic, and Formalization*, London and New York: Routledge.

—— (ed.) (1992c) *Proof and Knowledge in Mathematics*, London and New York: Routledge.

—— and Luker, M. (1980) 'The Four-colour theorem and Mathematical Proof', *Journal of Philosophy* 803–20.

Devlin, B. *et al.* (1997) *Intelligence, Genes, and Success: Scientists Respond to* The Bell Curve, New York: Springer.

Devlin, K. (1977) *The Axiom of Constructability*, Berlin: Springer.

Diacu, F. (1996) 'The Solution of the *n*-body Problem', *The Mathematical Intelligencer* 18(3): 66–70.

Dieudonné, J. (1969) *Foundations of Analysis*, New York: Academic Press.

Drake, F. (1974) *Set Theory: An Introduction to Large Cardinals*, Amsterdam: North Holland.

Duhem, P. (1906) *The Aim and Structure of Physical Theory*, Princeton: Princeton University Press.

Dummett, M. (1959a) 'Wittgenstein's Philosophy of Mathematics', reprinted in Dummett (1978b).

—— (1959b) 'Truth', reprinted in Dummett (1978b).

—— (1973) 'The Philosophical Basis of Intuitionistic Logic', reprinted in Dummett (1978b).

—— (1973) *Elements of Intuitionism*, Oxford: Oxford University Press.

—— (1978a) 'Frege's Distinction Between Meaning and Reference', reprinted in Dummett (1978b).

—— (1978b) *Truth and Other Enigmas*, Cambridge, MA: Harvard University Press.

—— (1991) *Frege: Philosophy of Mathematics*, Cambridge, MA: Harvard University Press.

—— (1991a) *The Logical Basis of Metaphysics*, Cambridge, MA: Harvard University Press.

Dunham, W. (1994) *The Mathematical Universe*, New York: Wiley.

Edwards, H.M. (1974) *Riemann's Zeta Function*, New York and London: Academic Press.

Enderton, H. (1977) *Elements of Set Theory*, New York: Academic Press.

Epstein, D. *et al.* (1992) 'About This Journal', *Experimental Mathematics* 1(1).

Ewald, W. (1996) (ed.) *From Kant to Hilbert: A Source Book in the Foundations of Mathematics*, 2 vols., Oxford: Oxford University Press.

Fallis, D. (1996) 'Mathematical Proof and the Reliablity of DNA Evidence', *American Mathematical Monthly* 103: 491–97.

—— (1997) 'The Epistemic Status of Probabilistic Proof', *Journal of Philosophy*, 165–86.

Field, H. (1980) *Science Without Numbers*, Princeton: Princeton University Press.

—— (1989) *Realism, Mathematics, and Modality*, Oxford: Blackwell.

Folina, J. (1992) *Poincaré and the Philosophy of Mathematics*, New York: Palgrave.

Francis, G.K. (1987) *A Topological Picture Book*, New York: Springer-Verlag.

Franks, J. (1989) Review of Gleick, *Chaos: Making a New Science* (with reply to Gleick's reply), *Mathematical Intelligencer* 11(1).

Frege, G. (1884) *Foundations of Arithmetic*, trans. J.L. Austin, Oxford: Blackwell, 2nd edition 1953.

—— (1892) 'On Sense and Reference', trans. and reprinted in Geach and Black, *Translations from the Philosophical Writings of Gottlob Frege*, Oxford: Blackwell, 1970.

—— (1918) 'The Thought: A Logical Inquiry', trans. A. Quinton and M. Quinton, reprinted in Klemke (1968).

—— (1971) *On the Foundations of Geometry and Formal Theories of Arithmetic*, trans. and ed. by E.H. Kluge), New Haven: Yale University Press.

—— (1979) *Posthumous Writings*, eds H. Hermes *et al.*, Oxford: Blackwell.

—— (1980) *Philosophical and Mathematical Correspondence*, eds G. Gabriel *et al.*, Oxford: Blackwell.

Freiling, C. (1986) 'Axioms of Symmetry: Throwing Darts at the Real Number Line', *Journal of Symbolic Logic* 51: 190–200.

Friedman, M. (1983) *Foundations of Spacetime Theories*, Princeton: Princeton University Press.

Galilo (*Discoursi*) *Two New Sciences,* trans. by S. Drake, Madison: University of Wisconsin Press, 1974.

Gary, M. and Johnson D. (1978) *Computers and Intractability: A Guide to the Theory of NP-Completeness*, New York: Freeman.

Gautier, Y. (1991) *La Logique interne*, Paris: Vrin.

—— (1994) 'Hilbert and the Internal Logic of Mathematics', *Synthese* 101: 1–14.

George, A. and Velleman, D. (2002) *Philosophies of Mathematics*, Oxford: Blackwell.

Giaquinto, M. (1994) 'Epistemology of Visual Thinking in Elementary Real Analysis', *British Journal for the Philosophy of Science.*

—— *The Search for Certainty*, Oxford: Oxford University Press.

—— (forthcoming) *Visual Thinking in Mathematics: An Epistemological Study*.

Gleick, J. (1987) *Chaos: Making a New Science*, New York: Viking Penguin.

—— (1989a) Reply to Franks (1989), *Mathematical Intelligencer*, 11(1).

—— (1989b) Reply to Hirsch (1989), *Mathematical Intelligencer*, 11(3).

Gödel, K. (1931) 'On Formally Undecidable Propositions of Principia Mathematica and Related Systems', reprinted in J. van Heijenoort (ed.), *From Frege to Gödel*, Cambridge, MA: Harvard University Press, 1971; also in Kurt Gödel, *Collected Works*, vol. I, Oxford: Oxford University Press, 1986.

—— (1944) 'Russell's Mathematical Logic', reprinted in P. Benacerraf and H. Putnam (eds.) *Philosophy of Mathematics*, Cambridge: Cambridge University Press.

—— (1947) 'What is Cantor's Continuum Problem?', reprinted in P. Benacerraf and H. Putnam (eds) *Philosophy of Mathematics*, Cambridge: Cambridge University Press.

Goldfarb, W. (1985) 'Kripke on Wittgenstein on Rules', *Journal of Philosophy.*

Goodman, N. (1965) *Fact, Fiction and Forecast*, Indianapolis: Bobbs-Merrill.

—— (1976) *Languages of Art,* Indianapolis: Hackett.

Gorenstein, D. (1979) 'The Classification of Finite Simple Groups', *Bulletin of the American Mathematical Society*, N.S. 1: 43–199.

——, Lyons, R. and Solomon, R. (1994) *The Classification of Finite Simple Groups*, Providence, RI: American Mathematical Society. (This is the first of a projected dozen volumes. It contains a history and an overview of the general proof.)

Hahn, H. (1956) 'The Crisis in Intuition', in J.R. Newman (ed.), *The World of Mathematics,* New York: Simon and Schuster.

Hale, B. (1987) *Abstract Objects*, Oxford: Blackwell.

Hale, B. and Wright, C. (2001) *The Reason's Proper Study*, Oxford: Oxford University Press.

Hallett, M. (1984) *Cantorian Set Theory and Limitation of Size*, Oxford: Oxford University Press.

—— (1990) 'Physicalism, Reductionism, and Hilbert', in Irvine (1990).

—— (1994) 'Hilbert's Axiomatic Method and the Laws of Thought', in George (1994).

—— (1995) 'Hilbert and Logic', in M. Marion and R.S. Cohen (eds), *Québec Studies in the Philosophy of Science*, vol. I, Dordrecht: Kluwer.

Hammer, E. (1995) *Logic and Visual Information*, Stanford: CSLI.

Hardy, G.H. (1929) 'Mathematical Proof', *Mind* 38:

Hellman, G. (1989) *Mathematics Without Numbers*, Oxford: Oxford University Press.

—— (1993a) 'Gleason's Theorem is Not Constructively Provable', *Journal of Philosophical Logic* 22: 93–203.

—— (1993b) 'Constructive Mathematics and Quantum Mechanics', *Journal of Philosophical Logic* 22: 221–48.

—— (1993c) 'On the Scope and Force of Indispensability Arguments', D. Hull *et al. PSA 1992*, 456–64.

Helmholtz, H. (1887) 'An Epistemological Analysis of Counting and Measurement', in R. Kahl (ed.), *Selected Writings of Herman von Helmholtz*, Middelton, CT: Wesleyan University Press, 1971.

Herrnstein, R. and Murray C. (1996) *The Bell Curve: Intelligence and Class Structure in American Life*, New York: Simon and Schuster.

Heyting, A. (1956) *Introduction to Intuitionism*, Amsterdam: North Holland.

Hilbert, D. (1899) *Foundations of Geometry* (10th edition, trans. L. Unger), La Salle, IL: Open Court.

—— (1925) 'On the Infinite', reprinted in Benacerraf and Putnam (1983).

Hirsch, M. (1989) 'Chaos, Rigor, and Hype', *Mathematical Intelligencer*, 11(3).

Horgan, J. (1993) 'The Death of Proof', *Scientific American*, October:

Huntington, S. (1998) *The Clash of Civilizations and the Remaking of World Order*, New York: Simon and Schuster.

Irvine, A. (1989) 'Epistemic Logicism and Russell's Regressive Method', *Philosophical Studies*.

—— (ed.) (1990) *Physicalism in Mathematics*, Dordrecht: Kluwer.

Jacquette, D. (2002) (ed.) *Philosophy of Mathematics: An Anthology*, Oxford: Blackwell.

Jaffe, A. and Quinn, F. (1993) ' "Theoretical Mathematics": Toward a Cultural Synthesis of Mathematics and Theoretical Physics', *Bulletin of the American Mathematical Society* 29(1): 1–13.

Jones, V. (1990) 'Knot Theory and Statistical Mechanics', *Scientific American,* November:

Jubien, M. (1977) 'Ontology and Mathematical Truth', *Nous*.

Kari, L. (1997) 'DNA Computing: Arrival of Biological Mathematics', *Mathematical Intelligencer*, 9–22.

Kitcher, P. (1975) 'Bolzano's Ideal of Algebraic Analysis', *Studies in the History and Philosophy of Science*, 229–71.

—— (1976) 'Hilbert's Epistemology', *Philosophy of Science.*

—— (1983) *The Nature of Mathematical Knowledge*, Oxford: Oxford University Press.

Klee, V. and Wagon, S. (1991) *Old and New Unsolved Problems in Plane Geometry and Number Theory*, Washington: Mathematical Association of America.

Klemke, E.D. (1968) *Essays on Frege*, Urbana: University of Chicago Press.

Kline, M. (1972) *Mathematical Thought From Ancient to Modern Times*, Oxford: Oxford University Press.

Koblitz, N. (1988) 'A Tale of Three Equations; Or The Emperor Has No Clothes' and 'Reply', *Mathematical Intelligencer.*

Körner, S. (1960) *Philosophy of Mathematics: An Introduction*, New York: Harper and Row.

Krantz, D.H., Luce, R.D. Suppes, P. and Tversky, A. (1971–1990) *Foundations of Measurement* (3 vols.), New York: Academic Press.

Krantz, S. (1989) 'Fractal Geometry' (and reply to Mandelbrot 1989b), *Mathematical Intelligencer* 11(4).

Kreisel, G. (1967) 'Informal Rigour and Completeness Proofs', in I. Lakatos (ed.), *Problems in the Philosophy of Mathematics*, Amsterdam: North Holland, 138–71.

—— (1971) 'Observations on Popular Discussions of Foundations', in Scott (1971), 189–98.

Kripke, S. (1982) *Wittgenstein on Rules and Private Language*, Cambridge, MA: Harvard University Press.

Kunen, K. (1980) *Set Theory: An Introduction to the Independence Proofs*, Amsterdam: North Holland.

Lagrange, J.L. (1788) *Mécanique Analytic*, Paris.

Lakatos, I. (1976) *Proofs and Refutations*, Cambridge: Cambridge University Press.

Lakoff, G. and Núñez, R.E. (2000) *Where Mathematics Comes From*, New York: Basic Books.

Lam, C.W.H. (1990) 'How Reliable is a Computer-Based Proof?', *Mathematical Intelligencer* 12: 8–12.

—— (1991) 'The Search for a Finite Projective Plane of Order 10', *American Mathematical Monthly* 98: 305–18.

——, Thiel, L. and Swiercz, S. (1989) 'The Non-Existence of Finite Projective Planes of Order 10', *Canadian Journal of Mathematics* XLI, No. 6: 1117–23.

Lehman, H. (1979) *Introduction to the Philosophy of Mathematics*, Oxford: Blackwell.

Lickorish, W.B.R. and Millett, K. (1988) 'The New Polynomial Invariants of Knots and Links', *Mathematics Magazine* 61: 3–23.

Linsky, B. and Zalta, E. (1995), 'Naturalized Platonism vs Platonism Naturalized', *Journal of Philosophy*, XCII/10, October 1995.

Littlewood, J.E. (1953/1986) *Littlewood's Miscellany* (ed. B. Bollobás), Cambridge: Cambridge University Press.

Livingston, C. (1993) *Knot Theory*, Washington: Mathematical Association of America.

Lucas, J. (1961) 'Minds, Machines, and Gödel', in A.R. Anderson (ed.), *Minds and Machines*.

Maddy, P. (1990) *Realism in Mathematics*, Oxford: Oxford University Press.

—— (1997) *Naturalism in Mathematics*, Oxford: Oxford University Press.

Maddy, P. (2007) *Second Philosophy; A Naturalistic Method*, Oxford: Oxford University Press.

Malament, D. (1982) Review of Field, *Science without Numbers*, *Journal of Philosophy* LXXIX, 9: 523–34.

Mancosu, P. (1996) *Philosophy of Mathematics and Mathematical Practice in the Seventeenth Century*, Oxford: Oxford University Press.

Mancosu, P. (ed.) (1998) *From Brouwer to Hilbert*, Oxford: Oxford University Press.

Mandelbrot, B. (1977/1983) *The Fractal Geometry of Nature*, 2nd edition, New York: Freeman.

—— (1989a) 'Chaos, Bourbaki, and Poincaré', *Mathematical Intelligencer* 11(3)

—— (1989b) 'Some "Facts" That Evaporate Upon Examination', *Mathematical Intelligencer* 11(4)

Manders, K. (unpublished ms) 'Representational Granularity in the Philosophy of Mathematics'.

McDowell, J. (1992) 'Meaning and Intentionality in Wittgenstein's Later Philosophy', in P.A. French *et al.* (eds), *Midwest Studies in Philosophy*, vol. XVII, Notre Dame: University of Notre Dame Press.

Mendelson, E. (1987) *Introduction to Mathematical Logic*, 3rd edition, New York: Wadsworth.

Misner, C., Thorne, K. and Wheeler, J. (1973) *Gravitation*, Chicago: Freeman.

Monk, R. (1990) *Ludwig Wittgenstein: The Duties of Genius*, Jonathan Cape: London.

Moore, A.W. (1990) *The Infinite*, London: Routledge.

Moore, G. (1982) *Zermelo's Axiom of Choice: Its Origins, Development, and Influence*, New York: Springer.

Moravcsik, J. (1992) *Plato and Platonism*, Oxford: Blackwell.

Moschovakis, Y. (1990) 'Sense and Reference as Algorithm and Value', *Proceedings of the 1990 ASL Meeting*, preprint.

Mumford, D. (2000) 'Dawning of the Age of Stochasticity', in V. Arnold (ed.), *Mathematics: Frontiers and Perspectives*, New York: American Mathematical Society.

Mundy, B. (1987) 'Faithful Representation, Physical Extensive Meaurement Theory and Archinedean Axioms', *Synthèse*, 373–400.

Nagel, E. (1932) 'Measurements', *Erkenntnis*, 313–33.

Needham, T. (1997) *Visual Complex Analysis*, Oxford: Oxford University Press.

Nelsen, R. (1993) *Proofs Without Words*, Washington: Mathematical Association of America.

Nelsen, R. (2000) *Proofs Without Words* II, Washington: Mathematical Association of America.

Norman, J. (2006) *After Euclid: Visual Reasoning and the Epistomology of Diagrams*, Stanford: CSLI.

Papineau, D. (1993) *Philosophical Naturalism*, Oxford: Blackwell.

Parsons, C. (1979/1980) 'Mathematical Intuitionism', *Aristotelian Society Proceedings*.

—— (1983) *Mathematics in Philosophy*, Ithaca, NY: Cornell University Press.

—— (1990) 'The Structuralist View of Mathematical Objects', *Synthèse* 84: 303–46.

Penrose, R. (1989) *The Emperor's New Mind*, Oxford: Oxford University Press.

—— (1994) *Shadows of the Mind*, Oxford: Oxford University Press.

Pierce, C.S. (1957) 'On the Reality of Thirdness', in V. Thomas (ed.), *Essays in the Philosophy of Science*, Indianapolis: Bobbs-Merrill.

Pitcher, G. (ed.) (1966) *Wittgenstein*, Garden City, NY: Doubleday.

Pollard, S. (1990) *Philosophical Introduction to Set Theory*, Notre Dame: Notre Dame University Press.

Polya, G. (1954) *Mathematics and Plausible Reasoning*, Princeton: Princeton University Press.

Popper, K. (1972) *Objective Knowledge*, Oxford: Oxford University Press.

Pour-El, M. and Richards, I. (1983) 'Noncomputability in Analysis and Physics: A Complete Determination of the Class of Noncomputable Linear Operators', *Advances in Mathematics* 48(1): 44–74.

Putnam, H. (1971) *The Philosophy of Logic,* New York: Harper and Row. (Also reprinted as a chapter in the author's *Philosophical Papers*, vol. I, 2nd edition.)

—— (1975) 'What is Mathematical Truth?', *Mathematics, Matter and Method: Philosophical Papers*, vol. I, Cambridge: Cambridge University Press.

—— (1994) Review of Penrose, *Shadows of the Mind*, *New York Times Book Review*, November 20.

Quine, W.V. (1936) 'Truth by Convention', reprinted in Benacerraf and Putnam (1983).

—— (1951) 'Two Dogmas of Empiricism', reprinted in *From a Logical Point of View*, Cambridge, MA: Harvard University Press.

—— (1960) *Word and Object*, Cambridge, MA: Harvard University Press.

—— (1970) *Philosophy of Logic*, Englewood Cliffs, NJ: Prentice-Hall.

Reichenbach, H. (1938) *Experience and Prediction*, Chicago: Chicago University Press.

Reid, C. (1970) *Hilbert*, New York: Springer.

Resnik, M. (1980) *Frege and the Philosophy of Mathematics*, Ithaca: Cornell University Press.

—— (1981) 'Mathematics as a Science of Patterns: Ontology and Reference', *Nous* 15:

—— (1982) 'Mathematics as a Science of Patterns: Epistemology', *Nous* 16: 95–105.

—— (1988) 'Mathematics from the Structural Point of View', *Revue Internationale de Philosophie* 42: 400–24.

—— (1997) *Mathematics as a Science of Patterns*, Oxford: Oxford University Press.

Ribenboim, P. (1988) *The Book of Prime Records*, New York: Springer.

Riemann, G. (1859) 'On the Number of Primes Less than a Given Magnitude', trans. and reprinted in Edwards (1974).

Robinson, A. (1964) 'Formalism 64'.

Robinson, R. (1950) *Definition*, London: Oxford University Press.

Russell, B. (1903/1937) *The Principles of Mathematics*, Cambridge: Cambridge University Press.

—— (1907) 'The Regressive Method of Discovering the Premises of Mathematics', reprinted in D. Lacky (ed.) *Essays in Analysis*, New York: George Braziller.

Saaty, T.L. and Kainen, P.C. (1986) *The Four-Color Problem*, New York: Dover.

Sagan, H. (1994) *Space-Filling Curves*, New York: Springer.

Savage, C.W. and Ehrlich, P. (eds) (1992) *Philosophical and Foundational Issues in Measurement Theory*, Hillsdale, NJ: Lawrence Erlbaum.

Scott, D.S. (ed.) (1971) *Axiomatic Set Theory*, Providence, RI: American Mathematical Society.

Senechal, M. (1998) 'The Continuing Silence of Bourbaki – An Interview With Pierre Cartier', *Mathematical Intelligencer* 20(1): 22–28.

Shanker, S. (1987) *Wittgenstein and the Turning-Point in the Philosophy of Mathematics*, New York: SUNY Press.

Shanks, D. (1993) *Solved and Unsolved Problems in Number Theory,* 4th edition, New York: Chelsea.

Shapiro, S. (1983a) 'Conservativeness and Incompleteness', *Journal of Philosophy* LXXX, 19: 521–31.

—— (1983b) 'Mathematics and Reality', *Philosophy of Science* 50(4): 523–48.

—— (1989) 'Structure and Ontology', *Philosophical Topics*.

—— (1997) *Philosophy of Mathematics: Structure and Ontology*, Oxford: Oxford University Press.

Shapiro, S. (2000) *Thinking About Mathematics*, Oxford: Oxford University Press.

Shin, Sun Joo (1994) *The Logical Status of Diagrams*, Cambridge: Cambridge University Press.

Sieg, W. (1999) 'Hilbert's Programs: 1917–1922', *The Bulletin of Symbolic Logic* 5(1): –44.

Stein, H. (1990) 'Eudoxos and Dedekind: On the Ancient Greek Theory of Ratios and Its Relation to Modern Mathematics', *Synthèse* 84: 163–211.

Steiner, M. (1975) *Mathematical Knowledge*, Ithaca, NY: Cornell University Press.

—— (1983) 'Mathematical Realism', *Nous* 363–84.

—— (1989) 'The Application of Mathematics to Natural Science', *Journal of Philosophy* LXXXVI, 9: 449–80.

—— (1998) *The Applicability of Mathematics as a Philosophical Problem*, Cambridge, MA: Harvard University Press.

Stroud, B. (1965) 'Wittgenstein and Logical Necessity', reprinted in Pitcher (1966).

Suppes, P. (1957) *Introduction to Logic*, New York: Van Nostrand.

Swoyer, C. (1987) 'The Metaphysics of Measurement', in J. Forge (ed.), *Measurement, Realism, and Objectivity*, Dordrecht: Reidel.

Tait, W.W. (1983) 'Against Intuitionism: Constructive Mathematics is Part of Classical Mathematics', *Journal of Philosophical Logic* 12: 173–95.

—— (1986a) 'Truth and Proof: The Platonism of Mathematics', *Synthèse*, 341–70.

—— (1986b) 'Plato's Second Best Method', *Review of Metaphysics*, 455–82.

Teller, P. (1980) 'Computer Proof', *Journal of Philosophy*, 797–803.

Thurston, W. (1994) 'On Proof and Progress in Mathematics', *Bulletin of the American Mathematical Society* 30(2): 161–77.

Tiles, M. (1984) 'Mathematics: The Language of Science?' *The Monist* 67(1): 3–17.

—— (1989) *The Philosophy of Set Theory*, Oxford: Blackwell.

—— (1991) *Mathematics and the Image of Reason*, London: Routledge.

Trudeau, R.J. (1976) *Introduction to Graph Theory*, reprinted 1993, New York: Dover.

Tymoczko, T. (1979) 'The Four-Color Problem and Its Philosophical Significance', *Journal of Philosophy*, 57–83.

Urquhart, A. (1990) 'The Logic of Physical Theory', in Irvine (1990).

Van Fraassen, B. (1980) *The Scientific Image*, Oxford: Oxford University Press.

Vesley, R. (preprint) 'Boolos' Nonconstructive Proof of Gödel's Theorem'.

Wagon, S. (1985a) 'The Evidence: Perfect Numbers', *The Mathematical Intelligencer* 7(2): 66–68.

—— (1985b) 'The Evidence: Is π Normal?', *The Mathematical Intelligencer* 7(3): 65–67.

—— (1986) 'The Evidence: Where are the Zeros of Zeta of s?', *The Mathematical Intelligencer* 8(4): 57–62.

Wang, H. (1974) *From Mathematics to Philosophy*, London: Routledge and Kegan Paul.

—— (1987) *Reflections of Kurt Gödel*, Cambridge, MA: MIT Press.

—— (1996) *A Logical Journey*, Cambridge, MA: MIT Press.

Wang, Q. (1991) 'The Global Solution of the n-body Problem', *Celestial Mechanics* 50: 73–88.

Weyl, H. (1918) *The Continuum*, trans. Pollard and Bole, New York: Dover.

—— (1949) *Philosophy of Mathematics and Natural Science*, Princeton: Princeton University Press.

Whitehead, A.N. and Russell, B. (1910) *Principia Mathematica*, Cambridge: Cambridge University Press.

Wigner, E. (1960) 'The Unreasonable Effectiveness of Mathematics in the Natural Sciences', reprinted in *Symmetries and Reflections*, Cambridge, MA: MIT Press, 1967.

Witten, E. (1989) 'Quantum Field Theory and the Jones Polynomial', *Communications in Mathematical Physics* 121: 351.

Wittgenstein, L. (1922/1961) *Tractatus Logico-Philosophicus*, London: Routledge.

—— (1953) *Philosophical Investigations*, eds G.E.M. Anscombe and R. Rhees, Oxford: Blackwell.

—— (1956/1978) *Remarks on the Foundations of Mathematics*, eds G.H. von Wright, G.E.M. Anscombe and R. Rhees, Oxford: Blackwell.

—— (1989) *Wittgenstein's Lectures on the Foundations of Mathematics, Cambridge 1933*, ed. C. Diamond, Chicago: University of Chicago Press.

Woodin, W. Hugh (2001) 'The Continuum Hypothesis, Parts I and II', *Notices of the American Mathematical Society*, June/July, 567–76, and August, 681–90.

Wright, C. (1980) *Wittgenstein's Philosophy of Mathematics*, London: Duckworth.

—— (1983) *Frege's Conception of Numbers and Objects*, Aberdeen: Aberdeen University Press.

—— (1993) *Realism, Meaning, and Truth*, Oxford: Blackwell.

Zwicky, J. (2003) *Wisdom and Metaphor*, Kentuille, NS: Gaspereau Press.

Index

Compiled by Mary Leng (first edition) and revised by Devin Suderman (Second edition)